Change Your World

Praise for *Change Your World*

"Michael Ungar's Change Your World shows that recovery, functioning and positive change in the face of adversity is not a lonely path trod by individuals; here lies the personal and social transformative power of resilience."
— Joel Reyes, Sr. Education and Institutional Development Specialist, World Bank

"Ungar's intelligent, bracing, and frequently moving narrative shows us that genuine resilience lies not in 'what we have, but what we receive'— a lesson that should prompt a major rethink at almost every level about what it means to live and thrive in today's complex globalised world."
— Michele Grossman, Research Chair in Diversity and Community Resilience, Deakin University, Australia

"Combining solid research with inspiring stories, Ungar elegantly guides us toward what really creates change and hope: meaningful relationships, consequential resources, and a social context that supports thriving and respect. *Vive la résilience!*"
— Frank N. Thomas, Professor of Counseling & Counselor Education, Texas Christian University

"Well-written and deeply researched."
— Kirkus Reviews

"Ungar's approach makes [his] theory accessible, portraying it in human terms."
— Clarion Reviews

"Finally, a master story-teller presents a clear-eyed case for why the systems that surround us determine more where we emerge in life than our personal qualities. Working in a developing country, I am struck by the insistent narrative that we hold keys to our success in our hands. Try telling that to a child growing up in poverty, immersed in a failing school, and surrounded by a poor healthcare system."
— Dipak Naker, Cofounder and Codirector, Raising Voices

Change Your World

The Science of Resilience and the True Path to Success

Michael Ungar, PhD

sh.
SUTHERLAND
HOUSE

Sutherland House
416 Moore Ave., Suite 205
Toronto, ON M4G 109

Sutherland House and logo are registered
trademarks of The Sutherland House Inc.

Second edition, September 2019

If you are interested in inviting one of our authors to a live event or media appearance, please contact publicity@sutherlandhousebooks.com and visit our website at sutherlandhousebooks.com for more information about our authors and their schedules.

Manufactured in Canada
Cover designed by Sarah Beaudin
Cover photograph courtesy iStock
Book composed by Karl Hunt

Library and Archives Canada Cataloguing in Publication
Title: Change your world : the science of resilience
and the true path to success / by Dr. Michael Ungar.
Names: Ungar, Michael, 1963- author.
Description: Includes bibliographical references.
Identifiers: Canadiana 20190043695 | ISBN 9781999439521 (softcover)
Subjects: LCSH: Resilience (Personality trait) | LCSH: Success. |
LCSH: Self-actualization (Psychology) | LCSH: Interpersonal relations.
Classification: LCC BF698.35.R47 U54 2019 | DDC 155.2/4—dc23

ISBN 978-1-9994395-2-1

Contents

Chapter 1

Our Success
Depends on Others

AN ACQUAINTANCE OF MINE is the manager of a large coal mine just outside Johannesburg, South Africa. I know the mine, or at least the township where the workers live, their homes part of an informal settlement surrounding tall elevators that bring the coal to the surface. Snaking for miles beneath the town are deep shafts dug by generation upon generation of miners—shafts that occasionally collapse, leaving sinkholes in the streets above. Mountains of tailing rise like sand dunes on the horizon, their dry orange dust raised by yellow bulldozers that from a distance look like children's toys in a giant sandbox. The dust from the tailing coats everything and everyone when the rains don't come – and they rarely come. Shallow streams in the vicinity are filled with plastic bags and empty bottles.

Parents in the settlement tell their children to stay in school and get an education. What they mean by *education* is science. Forget the arts; forget reading for pleasure. A scientific education is the only way forward in a town that knows nothing but moving coal, the same coal that is heating the atmosphere and drying up the land. The same coal that will no longer be wanted sometime soon when environmentalists, not popular here, have their way. It's easy to understand why parents are so focused on preparing their children for a life underground, or in one of the extractive industries. There are few jobs that do not depend on the mine. Those who are lucky enough to have work save their money and dream of moving their families out from under the corrugated steel roofs of the informal settlements into brick homes with indoor toilets. Very few make it. Most will remain sleeping in mud-floor lean-tos under lampposts that reach to an impossible height. At night, these lampposts give the community the feeling of a prison without guards.

The lampposts are extra tall to prevent any unauthorized person from fussing with the wires. Some residents feel the risk of electrocution is a small price to pay for hooking themselves onto the power grid for free. They are also high because young men with nothing but time on their hands break the bulbs with rocks to prove they are as powerful as their warrior ancestors or the modern-day superheroes they watch online. Darkness is the reward for their foolishness. One cannot blame them, really. As wages have risen, coal production has been automated, and mines are not hiring. The workers kept on are older men with the seniority and technical expertise required for efficient operations in a dangerous environment, miles beneath the earth.

Automation has also brought on a spike in workplace deaths, not because of unintended injuries but because older miners have a work ethic that puts them at risk for heart attacks and strokes. My friend, the mine manager, does not want to sound callous, but he sees the irony of the situation. The commitment of the older men to getting their jobs done efficiently actually jeopardizes the profitability of the mine. The old guys will bore and blast at the same pace as they did when they were younger, only now the stress on their bodies routinely causes them to overexert. Each heart attack or stroke causes a lengthy delay in production while the victim (or body) is brought to the surface and an investigation is carried out by authorities. Bribes have to be paid. Unions threaten to strike. There are grieving widows and bad press. It is a serious problem for everyone involved.

The solution the mine has put in place, however, is inspiring, and it tells us a lot about how we can best help people thrive. It was arrived at in stages. Initially, the mine managers provided the miners with a simple wearable device that monitored their heart rates and other physiological data. The system was designed so that an alarm would ping in an office at the surface if a worker's vitals were dangerously out of control. In other words, the workers' supervisors would know if one of their men was medically at risk and could send word down the mineshaft for the employee to take a break. The union reassured their members that there would be no loss in pay or other consequences if they took a few minutes to rest. That did not work: the workers kept breaking the monitors because (no surprise) they felt like they were being watched, and they worried that they would be fired if they could not work without interruption. Once the supervisors figured out their mistake, they went to the union and asked if the shop stewards would monitor employees' heart rates. As expected, the union did not want the responsibility or the liability, nor did it want to appear to be in bed with management. There was another solution, however, and it should have been obvious from the start: it came in the form of the men's wives, girlfriends, and mothers.

The company gave these women a small alarm that went off if their husband, boyfriend, or son's vitals were compromised. It was then up to the women to alert the team leaders underground that something was wrong. I was told that the men stopped breaking their monitors, mostly because they feared a confrontation with the people who loved them and to whom they felt responsible. While I question the ethics of burdening women with the responsibility to keep miners alive, the example nevertheless holds interesting lessons about what makes people resilient. The solution to heart attacks and strokes among the miners did not rely on an individual change in behavior. The mine's Human Resources department could have taught the miners to meditate, self-regulate, or eat healthier. Having met a few of those rock-hard personalities, I doubt this would have worked. What did work were the external supports that gave the miners what they needed to survive in a genuinely tough environment. Managers, unions, families, and even the physical infrastructure of the mine had to change to make it possible for men to take breaks when they needed them without the potential for negative consequences. Simply put, the miners changed their behavior and began acting responsibly when the environment around them forced them to change. Acting in their own best interest was a better choice than going home to an angry wife, girlfriend, or daughter.

I wonder what could be accomplished if the same amount of creative thinking was applied to the settlement's other problems of poor housing, youth unemployment, and garbage-strewn streams. Taller lampposts are a necessary adaptation to a bad situation, but real change requires more thought and a better understanding of the science of resilience.

★ ★ ★

I enjoy an inspiring TED Talk as much as anyone else. I love that "Ah ha!" moment when I gain some new insight into myself or, at the very least, better understand why everyone else is so dysfunctional. I understand why motivational speakers seek to inspire us with the message that no situation is so horrific that it cannot be solved by an individual's strength and will. When bad things happen and our lives hit a wall, therapists and self-help authors give us tales of exceptional individuals who not only survive adversity but thrive in toxic environments. Like everyone else, I want to believe that attainments as complex as success, happiness, love, and meaning can be attributed to a short list of personal traits like virtue ("I've always been a kind-hearted person"), faith ("I never stopped believing that my life had a higher purpose"), perseverance ("I never stopped trying to make things better"), self-control ("I waited until I knew the time was right to act"), resistance ("No one could ever tell me I wasn't good enough"), talent ("From a young age, I was very good at what I do"), optimism ("Life gets better if you let it"), and grit ("I never accepted that I was weak even though others told me I was"). If we just listen to the right podcasts or sign up for the right courses, we will discover our hidden strengths and find the happy life that is waiting to burst from inside us. Life, we are told, is up to the individual.

Sadly, these stories are misleading. Sometimes they are downright lies. Surely, by now, we all have some experience of the shortcomings of the self-help approach to life. We have fallen prey to the seduction of hope. We have made the resolutions and done the diets, the meditations, the exercises, and the positive thinking, but whatever changes we make endure for a short amount of time. We dutifully listen to the business gurus, consultants, and coaches in our workplaces. We take upon ourselves the task of becoming motivated. We subject ourselves to the heavy lifting of personal transformation even when our world remains stubbornly the same. We mostly fail: we gain back the weight that we lost; our next relationship is just as bad as the one we left; the neighborhood we relocate to is just as unfriendly as the old one. Self-help fixes are like empty calories: the effects are fleeting and often detrimental in the long term. This is because the stresses that put our lives in jeopardy in the first place remain in our environment.

The science of resilience shows us that our lives can get better, and most people's lives do get better, but not because of what motivational speakers are selling. It is true that as human beings, we have internal resources to help us thrive in the most emotionally and physically damaging situations, but those internal resources are seldom of much use unless we are also given the external resources we need to succeed. Here's the problem in a nutshell: we have fallen in love with the persona of the rugged individual. We need to give more consideration to the *resourced* individual. Rugged individuals make great television; resourced individuals make good science.

I have witnessed examples of personal success in economically secure countries and forgotten parts of the world where one expects to find only despair and failure. Regardless of where the stories are set, people always want to convince me that their success hinges on them being exceptional individuals with the right thoughts, feelings, behaviors, personalities, and genetics. If I stay a while, however, and get to know a person's life story, I soon discover the truth about their resilience. They might have fine personal qualities, but their success is dependent on the support they receive from the people around them. Resourced individuals do much better than individuals without resources, regardless of personal qualities. Miners do not avoid heart attacks when left to themselves; awful employers do not become pleasant bosses just because employees change their attitudes or perceptions of their workplace; and an emotionally abusive family life does not improve without spouses and children stepping up and taking responsibility for the part each plays in creating the problem.

Success depends upon our ability to change the world around us far more than our ability to change ourselves. This one straightforward observation has huge ramifications, not only for the self-help industry but for religion, medicine, and politics, as well. It suggests that our success is a product of how well others provide us with what we need to survive and thrive, and that motivation is only a small part of the reason some people do well and others do not.

Plenty of science exists to demonstrate that changing the world, rather than ourselves, is the best way to make us resilient. It is also the easiest way to ensure success.

For more than 20 years, I have been a family therapist working with hard-to-reach young people while also holding a research chair that has let me study resilience around the world. The Resilience Research Centre at Dalhousie University, which I founded and still lead, investigates why some people "beat the odds" and do far better than expected. Unlike many other research centers, the team I have put together is focused not on our personal traits but rather on our social and physical ecologies—the natural environments in which we live—and how these create well-resourced individuals who make success look easy.

Our studies, whether in a South African township, a rural community on the Canadian Prairies, or the slums of Beijing, are helping to define resilience as much more than our individual ability to cope with adversity. Instead, we have demonstrated that our individual ability to flourish under stress can usually be attributed to our capacity, first, to navigate to the resources necessary for success—whether a psychologically supportive relationship, stable and affordable housing, or a nice job—and, second, to negotiate for the resources that are meaningful to us.

What do I mean by meaningful resources? And how do they facilitate our resilience? Before I began my research on resilience, I used to puzzle over why the children and families with whom I worked showed such vastly different outcomes even when they received the same clinical support. Then I met a strong-willed 15-year old girl who was in jail for stealing a car and crashing it into a ditch while trying to outrun the police. She taught me a thing or two about resilience, concepts that I would go on to study and publish. Plenty of people had offered that girl help, but no one had been able to get her to behave. In a meeting room in a secure custody facility, I asked her why she had refused to work with an addiction counselor, a guidance counselor, or a psychiatrist. She stared at me a while and then shrugged her shoulders, slumping down on the couch in the family visitation room where we were meeting. I kept silent in hopes that she would see that my question was not some counselor's trap. I genuinely wanted to know. Finally, she looked me in the eye and said, "None of them really got me. I like being a little shit."

It might not sound like a revelation, but this was a profound moment in my career. I suddenly understood that stealing cars brought with it far more advantages than disadvantages, including a sense of personal efficacy, a powerful identity, and bragging rights. What alternative was I offering as her counselor? Emotional support and the chance to heal old wounds? Then what? If I stopped her from being a thief, and a reasonably good one at that, what would she be instead? Academically, she was far behind her peers. Socially, other delinquents were the only kids who liked her reckless, bossy ways. Once I understood this, the light turned on. Not only did my clients need to be offered supports: they needed these supports to have value to them. They needed to be able to navigate their way to the resources they needed while negotiating for those resources to be provided in ways that were meaningful to them. These dual processes of navigation and negotiation would become the foundations of a long and successful series of studies on resilience.

But that would come later. For the moment, I dropped the pretense of counseling the girl and searched instead for new ways to make a car thief feel important and remain popular with her peers. It was not easy until she got work at the mall and realized that bossing people around behind the counter in the food court and dealing with customers was something she was good at. With the money she earned and the ability to sell her friends large fries for the price of a small, she soon experienced a strong identity that compensated for not being behind the wheel of a stolen vehicle. While a job might not sound like a big deal to those of us who do not steal cars, to the girl's way of thinking, serving burgers at the mall made her feel successful without the risk of incarceration.

★ ★ ★

Since that moment, my study of resilience has shifted attention from disorder, disease, and dysfunction to the coping strategies we use when facing adversity. I built an international program of research that at one point included a five-country, six-year study with colleagues from New Zealand, South Africa, Colombia, and China who replicated the work of my team in Canada. Together, we examined how 13-to-24-year-olds with complex needs living in stressed environments (like economically depressed neighborhoods and homes with family violence) make use of the health and social services available to them, and whether their patterns of service use are associated with their resilience over time. The point of the study was to explore a seldom discussed aspect of resilience, the services we receive from health, social welfare, and educational systems, as well as the informal supports we sometimes need from our families and communities. Rather than focusing our attention on individual factors like grit or mindset, we wanted to understand whether an investment in *services* could be a better way to nurture well-being in suboptimal environments. Remarkably few studies to date have asked the obvious question: Does resilience depend on the services we receive?

For our sample, we purposefully selected adolescents and young adults who were using multiple services. These were young people needing special educational supports at the same time that they were under the supervision of a child welfare worker because of exposure to family violence. Or they were youth with severe mental health problems like attention deficit disorder and conduct disorder who were also under a probation order because they had been caught selling drugs or committing a violent act. Some of our participants had learning challenges, others anxiety disorders. Some were homeless because they had run away from abusive parents. When data collection was completed, we methodically churned out statistics.

Finally, on a warm spring day after years of work, a senior statistician on the team came to me with a one-page graphic representation of a structural equation model that focused on Canadian youth. The math was daunting, but what it showed was the relationship between risk exposure, resilience, and behavioral outcomes for almost five hundred young people, all of them facing serious challenges.[1] We later verified these results with over 7,000 young people around the world, but this was the first proof that let us say with certainty that resilience depends more on what we receive than what we have. There, amid the squiggles and text boxes, was evidence that young people do much better when they receive a weave of services delivered with consistency in culturally relevant ways at a time and place valued by them. These resources, more than individual talent or positive attitude, accounted for the difference between youths who did well and those who slid into drug addiction, truancy, and high-risk sexual activity.

I have to admit, the diagram made me tear up. It is geeky, I'm sure, to react emotionally to schematics, but we had managed to prove that resourced individuals do far better than rugged individuals. We also discovered that the reason why many young people who need help do not take advantage of the help that is offered (if, and when, it is provided) is because service providers seldom tailor their programs to the clients' needs. For example, we heard stories of guidance counselors at children's schools who insisted parents take time off from minimum wage jobs to attend case conferences because guidance counselors and psychometricians do not work evenings. It should come as no surprise that the most vulnerable families did not show up because they could not afford the lost time at work. It was their children, doubly disadvantaged by learning difficulties and poverty, who wound up untreated and who eventually dropped out of school.

There were many more findings of that nature. We learned that if kids were not responding to treatment, it was not the kids' fault but a failure of the services to meet children's needs. We learned that young people who reported being motivated to turn their lives around had consistent relationships with a worker who respected them; they did not shuffle around from service to service. Shape the right environment for a troubled child, and the child changes for the better. Put in front of a child the necessary help, and he or she will take advantage of it. This is true even with children who are not initially motivated to make something of their lives.

Counterintuitively, we learned that the kids with the most problems who received *fewer* services reported *more* resilience and *better* outcomes than relatively untroubled young people who had access to loads of help. Clearly, more services do not make children better. The quality of the services received is what counts most. That is a politically dangerous finding because it can be misunderstood by bureaucrats looking to save money. However, our research does not say, "Provide fewer services." It says, "Provide the right service, from the right people, in the right way." This is crucially important to our understanding of resilience.

There is no reason to doubt that our findings hold true for adults. Personal explanations for success are convenient ways to make us feel better about ourselves for a moment, but they actually set us up for failure. TED Talks and talk shows full of advice on what to eat, what to think, and how to be in relationships may be immensely popular, but they have done little to stem the growth in the number of prescriptions to treat depression, the number of sick days taken at work, and the money spent on divorce lawyers.

The numbers are downright depressing. While the rate of divorce in the US has declined from 2000 to 2016 (down from a peak of 4.0 per thousand adults to 3.2), it is still high by historical standards, and the more important story is that the rate of marriages has declined by 15% over the same period of time.[2] Many young people are giving up on matrimonial commitments. They find living together or living alone to be preferable to what they see as the inevitable divorce.

Beyond our relationship problems, we are also experiencing alarming increases in our rates of obesity. Studies that rely on assessments of individuals by medical professionals show that the number of overweight adults in the United States has jumped by a third since 2000 to 39.6%. Two-thirds of American adults are now either overweight or obese. Worse, over that same 18-year period, obesity among youth has increased from 13.9% to 18.5%, and all indications are that it will continue to climb.[3] In Canada, the percentage of overweight adults grew from 27.8% in 1985 to 33.6% in 2011.[4] Obesity rates saw a steeper increase, from 6.1% in 1985 to 18.3% in 2011.[5] Even when we rely on people's self-reported data, the trend is similar: in 2014, 61.8% of men and 46.2% of women categorized themselves as overweight or obese, up from 57.3% and 41.3%, respectively, in 2003.[6] Again, these trends are happening while we enjoy easy access to self-help books and shelves of information and advice on healthy eating, most of which, unfortunately, put responsibility for change on ourselves as individuals.

Given the rise in obesity, it comes as no surprise that the rate of heart disease is growing almost as quickly. No matter which type of heart problem we focus on—ischemic heart disease, angina, arrhythmia, atrial fibrillation, or complete heart failure—about 158,700 Canadian adults, or 6.1 per 1,000 adults aged 20 and older, received a new diagnosis of heart disease in 2012.[7] The news is worse for men, whose rate of heart disease is twice that of women. The only positive development is that we are not dying at the same rate that heart problems occur, largely because of better access to improved care and life-saving medications. In fact, researchers who study the long-term trends in the use of cardiovascular medications in Canada over the past two decades have found that their use has increased sharply, with related costs rising over 200%. We now spend nationally over $5 billion on heart medications alone, sure a sign as any that while we are surviving heart disease, we are not preventing its onset with our diets or any other self-care routines.

Switching the focus to mental health, the story is just as troubling. Among adults, the number of hospitalizations in the US for mental health and substance use problems increased by 12% between 2005 and 2014.[8] While that may seem high, especially given the abundance of self-help materials available online and in person, what is more astounding is that prescription drug use, especially antidepressants, has increased fivefold since 1988. Over 10% of American adults now take at least one of these medications.[9] Likewise, there has been a 6% increase in the reported use of psychotropic drugs by adolescents.[10]

Other countries are doing no better, even when they share the same access as Americans to the self-help industry. Almost 10% of Canadians sought some form of mental health treatment for a mood or anxiety disorder in 2010.[11] More specifically, mood disorder rates for men and women rose from 5.1% in 2003 to 7.5% in 2013. As if these numbers were not bad enough, the rate at which we are filling prescriptions for mental health problems increased 353% in the 20-year period between 1980 and 2000[12]—and that was before the spike in anxiety and mood disorders among Canadians since the turn of the millennium.

As one might guess, all of this unhealthiness is keeping us from our jobs. In Canada, workers are recording more and more days absent, which can only mean that employees are either more stressed or more lazy. Either way, the story is not good. According to Statistics Canada, workers took an average of six sick days in 1980. That number rose to 9.3 by 2011 and shows no sign of reversing.[13] The story is even worse for government employees, who take 65% more sick days than people working in the private sector.

Again, these trends have occurred while we have been bombarded with advice on how to change ourselves. If that cumulative advice was working at all, it would be reasonable to assume that, as a population, we would be getting healthier and living with less stress. The reason that we are not seeing this bump in good physical health or mental hygiene is that we have been giving people the wrong message. Resilience is not a do-it-yourself endeavor. Striving for personal transformation will not make us better when our families, workplaces, communities, health care providers, and governments provide us with insufficient care and support. People who receive the rich resources they require from their environments, people who experience these resilience-promoting spaces, even creating them when they must, are far more likely to find success than individuals with positive thoughts but a toxic social environment and no help. If we think again about those South African miners but take away the heart monitors, the employers, the union, and the wife, mother, or daughter, their work would likely kill them.

★ ★ ★

Cinderella is one of my favorite tales of personal triumph over adversity. A stepchild is subjected to the cruelty of an evil stepmother and evil stepsisters until a fairy godmother arrives to help her attend the prince's ball, changing her life forever. It is a delightful bedtime story, but what does it really teach? Why does Cinderella get her prince?

It certainly helps that Cinderella, a servant girl who lives beneath the stairs and spends her days in rags, is delicate of feature, with long and silky hair. It also helps that she possesses a gentle, playful spirit and a joyful temperament. She might even have an outstanding genetic endowment that helps her ward off childhood diseases that might disable her, disfigure her, or kill her. All of these traits are advantageous to a child like Cinderella, and they predispose her to succeed, but the question remains: How much of her success is accounted for by her individual qualities alone? If the science of resilience is right, the answer is very little.

Indeed, life only turns around for this abused child when a fairy godmother appears and transforms a pumpkin into a carriage and mice into horses, sews Cinderella a beautiful gown, and fashions her a pair of glass slippers. Cinderella succeeds because her fairy godmother knows a thing or two about seducing a prince. I hate to kill the magic, but anyone would do better with that kind of support.

Imagine, for a moment, the story of Cinderella without the fairy godmother. A young girl is traumatized by the loss of both her parents. We have to assume that she copes with her grief on her own, as there is no mention of outpatient child psychiatry services in the fairy tale. In fact, there is no mention of a school to attend or a social worker to check in on Cinderella. (The prince holds glamorous balls, but the royal family apparently spends nothing on the welfare of its subjects.) The girl is a child laborer, made to work long hours as a household servant. She is treated abysmally by her stepfamily; she is isolated, traumatized, and denied education. She is at terrible risk of physical, sexual, and emotional abuse. Her only social supports seem to be the birds and mice she befriends.

Large epidemiological studies of children in adverse circumstances would give a child like Cinderella extremely high odds on developing a range of pathologies, from addiction to obesity. She would be a likely victim of domestic violence. The injustice of watching her stepsisters benefit from the family's wealth would stoke her resentment and anger. If Cinderella is at all like the many young people I have seen in my clinical practice, she would reach the age of 15, tell her stepmother to "fuck off," and run away. She would not wind up in the palace in the arms of her prince. She would wind up someplace behind the palace, exchanging sex for a bed to sleep, turning to prostitution and drugs to survive.

The difference between Cinderella-the-princess and Cinderella-the-prostitute is the fairy godmother. This is entirely consistent with the social science research on underprivileged children: success has more to do with the qualities of the people and resources in their lives than it has to do with the genotypical and phenotypical characteristics of the children themselves. And what of individual motivation? Cinderella's stepsisters seem far more determined to chase the prince than she does. Motivation is not nearly as important as being in an environment rich in opportunities. Who would not want to catch a prince (or princess) if given half a chance by a fairy godmother? My experience around the world suggests that motivation is something that is given to us by our circumstances rather than some internal quality. A world that provides us with the right mix of resources motivates us to act in our best interests. It also holds us accountable when we are lazy or ungrateful and overlook the opportunities that are within our grasp.

The problem with Cinderella stories is that they make success seem to be all about the personal qualities of exceptional people. They epitomize the myth of the exceptional individual and conveniently overlook the deficits in the environments that surround their protagonists. The heroes and heroines look so beautiful, talented, and self-assured that we forget about the supports that help them escape terrible situations. The story of the resourced individual is hidden in the footnotes.

It is not just fairy tales that deceive us. Take a real-life example: Barrington Antonia Irving, Jr., in 2007 became the youngest person and the first African American to pilot a plane solo around the world. Even more remarkable than his solo flight, Irving built his plane himself, after graduating *magna cum laude* from an aeronautical science program. He also founded an educational nonprofit dedicated to empowering minority youth to pursue careers in aviation. He did all this despite being born in Kingston, Jamaica, in 1983 and growing up in Miami's inner city surrounded by crime, poverty, and failing schools. The talk shows and magazines love Irving. He fits the narrative of an exceptional individual drawing on inner strength and remarkable talent to beat the odds—at least, it appears that way until we take a closer look at the factors that contributed to Irving's success.

While Irving is indeed a bright and driven young man, deserving of the accolades he receives, his ability to direct his own fate is a small part of the explanation for his success. Again, the science tells us, in statistical language, that motivation and talent account for a small amount of the variance in human outcomes.[14] Sure enough, if you scrape away the pull-yourself-up-by-your-own-bootstraps rhetoric from Irving's story, you learn that he was 15 and not feeling terribly bright when he chanced upon a meeting with a Jamaican-born pilot in the bookstore his parents owned. The pilot invited Irving to fly with him the next day and so inspired him that years later, Irving declined a university scholarship to play football and decided instead to wash planes at a local airport and get his pilot's license. When he later decided to fly solo around the world, he looked for sponsors. After more than 50 rejection letters from potential donors, he realized that the only way he would achieve his dream was to build a plane himself. He successfully solicited all of the parts he needed from aircraft manufacturers until he had a working plane and $30 in his pocket—enough, it seems, to set off around the globe.

Irving insists that people can do anything if they set their mind to it, and he now spends his time inspiring young people to pursue their dreams. That is wonderful, except that Irving is not your run-of-the-mill ghetto kid. He is, to use Malcolm Gladwell's term, an "outlier."[15] Outliers have the advantage of unusually supportive environments. In Irving's case, there was a family that must have greatly valued learning if his parents ran a bookstore in an impoverished community. There was the Jamaican pilot who was kind enough to give Irving a ride in his plane. There were schools and, I am going to assume, a few decent teachers and football coaches along the way who were able to give Irving the education and training he needed to get into university. There was the employer who let him wash planes. I also assume there was a public transportation system that helped him get out to the airport. There were dozens of philanthropists who donated plane parts worth over $300,000. And, of course, there was a tidal wave of social justice advocates who made it possible for a young African American to realize such an audacious dream in the first place. Thirty years earlier, it is unlikely that Irving would have been able to do what he did no matter how ambitious he was.

Barrington Antonio Irving, Jr., is an exceptional case, and he got more from his opportunities than others might have managed, yet that he got anything at all is due to an environment rich in resources. The same can be said of any young person who succeeds despite what may look like an unpromising situation. Resilience is almost always more nurture than nature.

Each of the above examples of personal success has been hinting at a short list of resources that make us more resilient. Over the years and through many different studies, I have found that at least 12 resources recur in resilient lives around the world. There may be many more, but these have stood out in my research and clinical practice thus far.

1. Structure: We all do better when the world around us provides routines and expectations, whether that is showing up at work or walking the dog twice a day. During a crisis, structure is even more important, as it offers a buffer against chaos. It makes us feel like our lives are predictable.

2. Consequences: Making mistakes is a prerequisite for success. The consequences we suffer, however, must offer manageable opportunities to repair what we have done wrong and integrate what we have learned into future efforts. A facilitative environment that improves resilience holds us accountable for our actions.

3. Intimate and sustaining relationships: Having even one person who loves us unconditionally is an important foundation for resilience. Even if that person is not in our life at the moment, we can do just fine with the memory of having once been well-loved.

4. Lots of other relationships: We all need a clan, a tribe, an extended family, colleagues at work, or an online community in which we feel we are needed. With loneliness becoming a public health crisis in many high-income countries, these networks of relationships have become more important than ever.

5. A powerful identity: How we are seen by others is crucial to our sense of self-worth. Our identities are co-constructions. We can tell others whom we want to be, but a good identity always depends on how others see us and whether they value what we have to offer.

6. A sense of control: Whether one experiences personal efficacy or political efficacy, we all do better when we are given the opportunity to make decisions that affect our lives.

7. A sense of belonging, religious affiliation, spirituality, culture, and life purpose: The list of things that give us a sense of belonging is long and culturally nuanced. Regardless of where we feel connected, we are more likely to succeed (especially during a crisis) when we feel our life has a purpose and others depend on us as much as we depend on them.

8. Rights and responsibilities: It is very difficult to experience success unless we experience social justice. It is also important that we are given genuine responsibilities for our own and others' welfare.

9. Safety and support: Knowing our homes and communities are safe and have the right supports in place to help us find the resources we need to cope when problems occur is a crucial component of our environments. Without these, we would be constantly overwhelmed by stress.

10. Positive thinking: People who succeed have a positive future orientation that is grounded in a realistic assessment of the opportunities they have been given. If we have plenty of resources and are still unappreciative of the advantages we enjoy, positive thinking can help us see that our problems are more in our heads than in the external world.

11. Physical well-being: Our environments provide us with everything from affordable, healthy food to sidewalks that encourage us to walk instead of drive. The better our environment is at keeping us physically healthy, the easier it is to maintain a lifestyle that improves our capacity to succeed.

12. Financial well-being: A strong economy, fair taxation, and poverty reduction strategies can all make us financially successful and impervious to changing economic conditions. How much money is enough is a question of social norms and the stigma that comes when we do not fit in with those around us.

Rugged individuals are supposed to create all of these resources for themselves. I find no evidence that they can. Resourced individuals, on the other hand, find what they need in their environments. If one stops and really ponders the reasons both Cinderella and Barrington Antonio Irving succeeded, they were offered resource-rich environments capable of lifting them out of a crisis. Who would not do better with a fairy godmother who can turn a pumpkin into a golden carriage, or with millionaire philanthropists willing to fund our most outrageous dreams?[16]

Plenty of research explains how changing the toxicity of our social environment changes our individual development. Stevan Hobfoll, a social scientist at Rush University in Chicago, uses terms like "environmental load" and "allostasis"[17] to describe the burdens and stresses our families and communities place on us. Obviously, our sense of well-being is undermined when our lives are cesspools of physical violence, economic exploitation, emotional abuse, and urban decay. We do better in secure spaces where we can reasonably expect to benefit from our efforts at self-improvement. When the environmental load is too great, only a Herculean amount of individual effort and a freakish amount of good fortune will make someone successful. Even then, as with Irving, we will find that unusual supports and resources are always part of the story.

Bad governments, bad economies, bad workplaces, and bad families, coupled with elitist, racist, sexist, and homophobic social norms, keep people stuck in toxic situations. In practice, this means our resilience depends on the resilience of our social institutions and individual or collective action. We ask our families to help us to get an education, we form neighborhood associations to keep our streets plowed, or we unionize and strike for fairer wages and safer working conditions. Maybe we take to the streets to protest racism or urge our politicians to address climate change before our coastal cities flood. Perhaps we leave a bad relationship or change whom we socialize with after work. A daughter might take the awkward step of refusing to come to Sunday dinner unless her mother stops being emotionally abusive toward others at the table. Or maybe our actions are more trivial: we write a letter to get fair compensation from the airline that lost our luggage. Stories of resilience can be small or big, but they always result in the world being changed in ways that make it easier to find the resources we need to sustain our well-being.

Sometimes, unfortunately, it is indeed left up to us as individuals to make change happen. We make better choices about our diet or adopt a new exercise routine so that we feel healthier. We insist our employer give us a raise. We divorce when a marriage goes stale. Divorce courts, not surprisingly, are full of people who said "enough is enough" and decided to get out of a bad relationship before it destroyed them. The good news is that most will remarry within five years and report far happier lives as a result of their decision. These examples all show that individual change is possible, although it is always easier to change when the institutions around us help us to fully realize our potential for success.

Motivation is not useless in these situations. Richard Wiseman is a British psychologist who famously placed in the middle of a magazine a large block of bold text that read: "Tell the experimenter you have seen this and win £250."[18] He then had a succession of people flip through the magazine. People who thought of themselves as lucky tended to see the ad and collect the cash. Those who perceived themselves as unlucky did not put themselves in the game and went home with nothing. Wiseman determined that luck comes to people who are skilled at creating and noticing chance opportunities. Lucky people make lucky decisions by listening to their intuition, creating self-fulfilling prophecies, holding positive expectations of themselves and others, and adopting a resilient attitude that transforms bad luck into good. Fair enough, but how much of this good luck is self-authored, and how much is the result of resource-rich (or luck-rich) environments? No one gets lucky unless someone puts £250 in a magazine in the first place.

I prefer to think, as Seneca said, that luck is what happens when preparation meets opportunity. The more opportunity in the environment, the more rewards there are for preparation, and the better the resourced individual succeeds. If, however, the soil in which we are planting our hopes is exhausted of nutrients, there is little likelihood that we will achieve a fertile crop of positive outcomes. Even strong individuals and lucky people will fail, no matter how hard they try.

Personal virtues are also useful in human development. Chris Peterson and his colleagues conducted a study in more than 50 countries that identified at least six common human virtues: wisdom, courage, humanity, justice, temperance, and transcendence. Each of these was found to depend on 24 character strengths, including hope, zest. self-regulation, curiosity, bravery, and so on.[19] All of these are worthwhile qualities to study, and they can sometimes predict great outcomes. The problem is that we do not know much about which virtues and character strengths are important in which social situations. Does self-regulation really matter when you are being bullied? (Would it be so bad if you threw the first punch?) Should a doctor working a 30-hour shift in an emergency room after a natural disaster practice self-regulation? Is bravery the best strategy when there is gun violence in the streets or a shooter in your school? Is curiosity a good thing if you want to remain a part of a fundamentalist church that teaches creationism? While it is comforting to imagine the world united by a common set of human traits, I am not sure which personality strengths bring the most success unless I understand the context in which those strengths are used. I am inclined to believe that the best personality type is the one most attuned to the demands placed upon us by our social situations or environments.

Xinyin Chen from the University of Western Ontario and a number of Chinese collaborators conducted an intriguing study, in which they examined personality traits and their social desirability among elementary school children in rural China. In 1990, just before market economics unshackled Chinese entrepreneurs, a reticent, deferential child was the one most likely to enjoy academic and social success. That association was weakened by 1998, and by 2002, shyness had become a risk factor for peer rejection, school failure, and depression.[20] The change occurring in just over a decade illustrates the power of the environment in determining which individual qualities matter most to success. Far from a unified story of individual strengths overcoming environmental constraints, environmental conditions shape the expression of individual qualities and predict their association with positive developmental outcomes.

Perhaps the most insidious aspect of promises of individual salvation in the self-help aisle is that they make us feel bad about ourselves. We are told that if we are failing, it must be because we are not trying hard enough; we do not have the right character strengths; we do not have good personal habits. We are supposed to be inspired by the people around us who supposedly make a personal path to success look possible; instead, we feel we cannot measure up and become demoralized, even though those people are not what they seem. While there are always a fortunate few who have great families and great jobs, who contribute to their communities and run marathons, most of these people, in my clinical experience, only *look* like they are doing it all, and doing it on their own. Beneath the veneers of their perfect lives, they are struggling with the same problems that plague everyone else. The neighbor who listens to motivational tapes, ranks first in her tennis division, and sells all that real estate has lost intimacy with her husband. The top engineer at the plant who practices yoga and does not eat meat is an alcoholic. Your sister has a large, perfectly organized home and fashionably dressed children, but her weight swings like the met-ronome that sits on the new piano that she never has time to play.

I know there can be exceptions to this rule. Miracles do hap-pen. But the rare individuals whose lives are as great as they seem are invariably rich in resources. Show me a successful child from a bad home, and I will show you a fairy godmother in the guise of an aunt, neighbor, teacher, coach, minister, or social worker. Show me a happy adult living a meaningful life, and I will show you an environment packed with opportunity and an individual willing and able to take advantage of what she is given. Good things happen to us when good things are happening in the world around us.

Social ecologists, community psychologists, and social workers have all found that if the world is the same today as it was yesterday, individual changes are unlikely to occur, and, if they do, they are unlikely to endure, no matter how much we believe in ourselves or how hard we try. At work, we can transfer from one department to another, but if the corporate culture has not changed, we will likely wind up hating our new boss just as much as we hated the previous one. At home, we can try to be thoughtful and mindful, and eat organic food, but if an unsupportive spouse refuses to give us the time we need to look after ourselves, it will not be long before the Tibetan singing bowl and the juicer start gathering dust in a corner. We need a clean break from the mindset that places the responsibil-ity for self-actualization on an individual's shoulders—it is a misread of what the science tells us about what makes us successful. If we want to understand why some people succeed and others do not, and if we want to succeed ourselves, we will need far fewer motiva-tional gurus and much more help from the people in our families, our workplaces, our communities, and our society.

Chapter 2

The Many Ways
We Cope

WE HAVE KNOWN FOR more than half a century that three things are necessary before we can decide on a strategy to cope with a bad situation: (1) we need to understand the kind of stress we are experiencing; (2) we need to identify the meaning we attribute to our experience of the stress; and (3) we need some resources on hand to avoid being overwhelmed by the demands of the stress.[21] Whether or not we can meet those three necessities determines whether a specific life event will turn into a crisis or become an opportunity for change.[22]

The odd thing about resilience—being able to endure the stress of a crisis—is that people show unique patterns of coping. A bad job interview may be catastrophic for one new graduate, while for another it merits nothing more than a shrug. A new chair that improves the life of the person working in the cubicle to my right does nothing to make life better for the colleague on my left. Someone who can handle a divorce away from a violent partner might be devastated by a divorce from a loved partner who cheated in the relationship.

In and of itself, stress is not necessarily a problem to be fixed. The stress of raising kids imposes a structure on our lives. Many of us actually function better and more efficiently when we have demands on our time. Just ask any parent whose second shift is getting a brood of kids to sports practice, and one quickly discovers that with a manic family life comes the opportunity to hone effective coping strategies. Research suggests that the right amount of stress can actually improve our ability to succeed far into the future, just like the DPTP-Hib vaccine prevents newborns from contracting diphtheria, tetanus, pertussis, polio, and Hib disease.

How much stress is the right amount to provide a benefit? When investigators examine the "steeling effect" of stress exposure, a curvilinear pattern emerges. Cancer patients who experience a moderately life-threatening (stage 2) cancer diagnosis develop coping skills that can benefit them for years to come. This advantageous pattern of adaptation under stress appears over and over again in studies of resilience, which makes me wonder why we are so afraid of failure and hardship, when it is precisely these experiences— when manageable, and when we overcome them—that make up the foundation for the knowledge and skills we need to succeed later on.

Indeed, people who have been exposed to moderately stressful events in the past tend to do better in traumatic circumstances, whether those be war, sexual assault, or a difficult final exam. This is not because they have become rugged individualists. Rather, their capacity to withstand and learn from stressful experiences hinges on how well-resourced they are before and after stress occurs. For example, children whose lives are completely bubble-wrapped break down under the weight of expectations later in life. As a university professor, I can assure you this is true. The most common form of cheating these days is not plagiarism; it is parents writing their children's assignments for them. As crazy as that sounds, there is a generation of university students who have been so sheltered that they cannot deal with the requirements of their courses. (One strategy professors now use is to have students write a short essay during class time at the beginning of the semester. They can compare this writing sample with what the student hands in as his or her final paper. I have noticed that parents are better than their children at differentiating between *their*, *they're*, and *there*.)

An optimal environment is one that provides just enough stress to allow people to learn and grow, while not overwhelming them with challenges that are beyond their capacity. In my earlier work, I noted a "risk-takers advantage," which follows from exposure to manageable amounts of risk and responsibility.[23] What I missed was finding the tipping point between too little and too much stress, something my research group has addressed more recently. Following on the heels of the study I described in Chapter 1, my team carried out a separate three-year investigation of 800 young people (aged 11 to 22 years) living in communities on the lower rung of the economic ladder or in rural communities where economic opportunities were few and far between. We knew these kids would have stress in their lives, but we did not know how much was good for them or how much was too much.

Our results were at first confusing, even disappointing. The main effects of risk on resilience are not especially significant. But tease apart the population of young people who score high and low on resilience and some exciting, albeit troubling, patterns emerge. For example, not all of the young people came from equally difficult communities. Among those who came from communities that were socially disorganized (meaning they had more graffiti and street violence), the kids with the most resilience were the least delinquent and the most invested in getting an education. That makes complete sense. We expect that a child growing up in a difficult community who nevertheless finds the right resources will avoid criminal behaviors and attend school regularly.

Something quite different occurs for children in communities with few social problems. Kids who scored *low* on resilience in these less stressful communities were the least likely to be delinquent or truant, while those who scored *high* on resilience surprised us with higher levels of criminal behavior and disengagement from school. The latter had grown up in stable, well-resourced environments, and they reported high levels of peer support, plenty of self-esteem, a sense of personal power, and parents who watched them closely. Yet, they were the most likely to be delinquent, truant, or both. In other words, young people who have no discernible exposure to stress, given the resources they need for resilience, are likely to turn out much worse than kids who have had to cope with personal challenges. The same protective factors that help a child survive in a harsh environment are disadvantageous to a child living in a safe, well-resourced, and relatively stress-free world. It may be that self-esteem and personal empowerment turns kids who have never had to struggle into entitled brats. Inflating a child's sense of self-worth without giving him opportunities to cope with adversity or take responsibility for himself and others may create plenty of narcissism but little socially desirable behavior.

It is a troubling finding, but it adds empirical proof to the idea that overprotective parents are actually harming their children. This has been suspected for a decade now, as huge spikes in the rate of anxiety disorders among children are causing a flood of referrals to emergency rooms and an increased number of hospitalizations for otherwise physically healthy young people.[24] Indeed, Canadian hospitals have seen a 44% increase in emergency room visits and a 33% increase in hospitalizations for mental health problems (mostly mood disorders like depression and anxiety) among children over the past decade. American health care providers are seeing similar changes, with a 7% increase in emergency room visits from 2006 to 2015. Much of this change can be accounted for by the increasing rate of mental illness reported among young people.[25] It seems that the overly protected child, having experienced few challenges in life and having been told how special she is, runs the risk of growing up to become a spoiled princess who does not understand the value of hard work or playing by the rules.

That is the strange and complicated truth about resilience. Our level of risk exposure changes how we experience the resources that help us succeed. This phenomenon is called *differential impact*.[26] In different environments, something that is beneficial when risk is low may have a disproportionately negative influence on us when risk is high. It all depends on the quality and quantity of the risks we face and the internal and external resources we have to marshal a response. If you have ever had a flight cancelled and become stuck in a long line of displaced passengers, you will know exactly what I mean. Passengers in the line will have profoundly different experiences depending on how disruptive the delay is to their lives, and their gratitude toward the helpful gate agent who fixes their problems also varies profoundly. As a general rule, the more "at risk" we are—the more the cancellation messes with our plans—the more our experience of helpful or protective factors will influence our sense of well-being.

Unfortunately, the differential impact of a resource is often overlooked when we explain why someone succeeds. We give the victor all the credit and conveniently forget that with the right mix of external resources, almost anyone can cope under conditions of adversity. This is a profoundly important idea. It reminds us that a mildly stressed (or even traumatized) child or adult may have greater abilities than her unstressed peers but that without a facilitative environment, she may not use her latent talents. It is not just the availability of resources that matters; it is whether they are the right resources in the right amount at the right time to help us tackle our worst problems.[27]

The theory of differential impact is the mirror image of the theory of *differential susceptibility*, which has become a well-loved concept in fields like epigenetics and neurophysiology. As many studies now show, different aspects of our biology make us more or less vulnerable to the experiences we have with the world around us.[28] For example, someone with a gene that puts them at risk for attention deficit hyperactivity disorder may perform better than his peers in a supportive environment but perform worse than his peers when the environment is stressful. This same pattern of susceptibility can also be found at the level of our genes. In a study led by Dante Cicchetti of the University of Minnesota's Institute for Child Development, it was shown that among women from low-income households with a diagnosis of major depressive disorder, those with one very specific variation of a gene associated with the functioning of the stress response system were far more likely than their peers to benefit from psychotherapy.[29] In other words, a genetic variation that influences one's style of learning and memory, specifically the amount that individuals ruminate on problems and the speed at which they make decisions,[30] was predictive of how susceptible these women were to the efforts of their psychotherapists.

Findings like this unfortunately contribute to the mistaken belief that our genes determine how we respond to the treatments we receive for everything from depression to substance abuse. Delve a little deeper, and the science shows the opposite. Change environments early in life, and a genetic trait that is either an advantage or a disadvantage may never be expressed. Change environments later in life, and we may be able to bring out the best in people by matching resources to genes—for instance, by offering therapy tailored to a genetic profile.

* * *

Imagine the 12 resilience resources I named in Chapter 1 are like bars of LED lights on an equalizer. As the music plays, some spike while others remain subdued. The subdued lights may still be there, but at that moment they are less obvious to an untrained ear. Music is dull when there are too few layers, and so it is with resilience. The more bars of light (resources) we have, the more likely we are to do well when life dishes up its random crises. It can be tricky to get the right resources at the right time, in the right order. Fortunately, those bars of light form predictable patterns that can guide us to desirable outcomes under appropriate circumstances. Context is crucial in the experience of resilience. Different combinations of resources are useful at different times.[31]

Of the several ways in which we show resilience, recovery is by far the most often discussed. We talk about a "recovery movement" and extol the virtues of Alcoholics Anonymous and its many spin-offs as pathways to mental health. The pattern here is that bad things happen, we do poorly for a time (which is normal), and then we bounce back. Our dip in functioning can last for a day, a month, or even many years depending on the severity of the stress, the resources we have to fix the situation, and what it all means to us. Take this simple example of job loss. Employees who have the skills to find other work, who have money to help them weather a few months of unemployment, and who do not perceive their job loss as their fault are likely to wade through that crisis almost unscathed. Recovery can be much more difficult when we do not understand why we were fired, when the job loss is our fault, or when we are unable to find new work.

Although we think of recovery as a return to normal, few people are left unchanged by the stress they experience. In this sense, the concept of recovery is ontologically flawed. We can never really recover and become the same person we were before a crisis. We are transformed, hopefully for the better, by our experience.

We re-evaluate our priorities. We retrain. We look for new work. We move. We take up a hobby. We start drinking. We stop drinking. We commit to a spiritual path. The options are endless. Recovery usually means a new regime that eventually becomes our new normal.[32]

Our recovery is seldom of our own doing. We recover because we find the resources we need. Consider the family whose child tells his parents he is gay. These days, most families cope reasonably well because we have changed the meaning of what being gay is all about. We see far more acceptance throughout society of people who are gay, lesbian, bisexual, and transgender. And that means that "recovering" as a family (if that is even any longer an appropriate description of the experience) when a child comes out can be relatively uncomplicated, depending on how the family handles the news.[33] If it is handled badly, research shows a much higher likelihood that a young person will engage in self-destructive behaviors or commit suicide. If it is handled well, the family remains strong. By my estimation, recovery is 70% reliant on the environment—in this example, how well the family copes and the sources of support they receive from their community-at-large—and 30% reliant on individual characteristics.

The possibility of normal, healthy functioning after stress is well within our grasp, but only if the world around us helps us with our recovery and gives positive meaning to our experience. That means victims of domestic violence and bullying can function as well as anyone with the right supports, such as friends and institutional policies that change the way police enforce the law.[34] Victims of racism can persevere when they are not only protected from discrimination but efforts are also made by the majority culture to deconstruct its privilege and promote tolerance and cultural pluralism.[35] In each case, *recovery* is a misnomer. We do not want people to simply return to their previous level of functioning or adjust to intolerable situations. Recovery should mean transformation into something better.

People tend to show more resilience than vulnerability during difficult times because most of us enjoy reasonably stable and supportive homes.[36] This explains why every longitudinal study of people coping over time tends to repeat some version of the same finding: a majority of individuals always survives reasonably well, because we live in a world where our needs are adequately met. We have relationships, education, a job. We belong to a church, a mosque, a temple, or a synagogue. We have safe streets and dozens of other advantages to fall back on when bad things happen. In situations like these, we are experiencing a pattern called *minimal-impact resilience*.[37] We still experience problems (being successful or showing resilience will not forestall life's challenges), but our problems will be more like beestings than severed arms. The pain is fleeting, and the chances of returning to normal are high. There is nothing magical about why so many of us do well in these contexts: it is in the DNA of our communities and our relationships to survive.

We have all heard the adage, "an ounce of prevention is worth a pound of cure." Resilience is the same. Those hours spent at church, community bingos, and fundraisers add up to strong networks that can be relied upon when our neighborhoods are under threat. If one wants to be strong during a crisis, it is best to invest in others before the crisis occurs. Celebrate a co-worker's birthday. Meet your children's teachers. Extend your education, and get familiar with the latest technologies. Vote. Volunteer. Make a casserole for a neighbor. Be sexually responsive to your spouse. Make sure elderly parents have a living will and savings in the bank. Collectively, we are more resilient when we are prepared for the dangers that lie waiting for us and we have on hand the necessary resources to weather a period of adversity. There is cause for optimism here. We can make ourselves more resilient by making the world around us supportive.

The problem is often not the stress itself as much as the meanings we attach to our stress and the threatening changes it produces in our environment. As George Bonanno of Columbia University says, not everything bad that happens to us is necessarily a problem.[38] Bonanno, an expert on resilience and grief, prefers to qualify all descriptions of traumatic events with the adverb *potentially*, as in something is potentially traumatizing rather than certain to be traumatic. After all, we will not know if something is traumatic until we experience it.

What we label as traumatic, and how we experience bad things, whether a child's or a job loss, is based on social interactions with others. People tell us how we should feel and we listen to them. Our emotions are as much a collective experience as they are individual. If you lose your job and everyone tells you, "There was nothing you could do. Everyone's job is disappearing," you are far less likely to blame yourself for your misfortune. Of course, there will be those who torment victims of injustice by telling them they are responsible for their downfall. An employer hands a pink slip to an employee and says, "You just didn't have what it takes," ignoring the fact that the business was mismanaged for years. Exchanges like these change our experience of a job loss and make it traumatizing. When it comes to feeling bad because bad things happen, packaging is everything.

When a drunk driver killed Candace Lightner's 13-year old daughter, Cari, on May 3, 1980, in Fair Oaks, California, Candace was expected to succumb to the emotional pain and, like most of us would, withdraw from the world. Instead, she started a movement that grew to become Mothers Against Drunk Driving (MADD). MADD has had a tremendous impact on strengthening laws against drunk driving and changing people's perceptions of driving while impaired. I mean no disrespect, but would Lightner have managed to have that much influence in the world without the loss of her child? For some time now, we have understood that the crises in our lives are also opportunities for unimagined growth. While we do not wish tragedy on ourselves or others, when tragedy does occur, it is often a catalyst to use our talents to achieve great things. This pattern is called *post-traumatic growth.*

Judging by the parade of extraordinary individuals who tell their stories on television and YouTube, post-traumatic growth is far more common than we think.[39] These are the amazing people who look after children with physical and intellectual disabilities, finding themselves to be more capable than they ever thought possible before they became parents.[40] These are also the spouses of soldiers who become better communicators, more loving, and better at keeping their households running because they find themselves on their own when their spouses deploy overseas.[41] The list is endless. The American psychiatrist and Harvard professor George Vaillant described post-traumatic growth in the simplest of terms: "If one loses one's shoes, one's soles become harder and resistant to corns."[42] Of course, like all wondrous patterns of resilience, one can never forget that it is still better to live in a world where if we lose our shoes, there are ways to get another pair without having to walk barefoot for long.

★ ★ ★

To this point, I have played the optimist. When resources are plentiful, affirmative patterns of resilience emerge. We recover. We grow. We simply get by and return to behaviors that approach normal with barely a hiccup. But what happens when our jobs migrate overseas, and our credit card debt goes wildly out of control, spurred by usurious interest charges? Bad situations like this seldom affect individuals alone. In a weak economy, for example, young adults are more likely to leave their communities to find work, leaving elderly parents to fend for themselves. Neighbors, too, lose their livelihoods, and with it their ability to maintain their homes. We know it can be much tougher to express resilience when the supports around us dwindle. When there are no socially desirable ways to adapt, we choose less desirable coping strategies. We have no choice. In impossible situations, the human spirit will do its best to find a way to survive.

Avoidant behavior might look like vulnerability, but there is plenty of research showing that withdrawing physically and emotionally from a bad situation can be an effective way of coping if one's circumstances remain stubbornly unchanged over time. Sadly, abused children are experts at the avoidant coping pattern. Neglect a child long enough, and she will learn to never depend on anyone for physical or emotional support.[43] "Which is better?" reasons the child: "Keep expecting to be loved and suffer disappointment, or simply hide from love and learn to rely entirely on myself for what I need?" A strategy like that can be incredibly useful, though I find it disheartening when I meet children who resort to it.

These avoidant strategies can be highly functional for years. When the American military was trying to increase resilience on the battlefield and avoid post-traumatic stress in its soldiers, it discovered a curious pattern. Soldiers who had been neglected and abused as children and who had adjusted by becoming emotionally distant from their parents were better able to withstand the stress of deployment in the unpredictable theaters of war in Afghanistan and Iraq.[44] In the abusive home and the theater of war, one never knows whom to trust. Avoidant strategies can work well in the right context. Of course, when children are adopted into loving homes, and soldiers with histories of abuse leave the military and try to form loving relationships, avoidant coping strategies can be a significant barrier to future happiness.

Other examples of the avoidant pattern abound. Acculturation by immigrants, for example, actually puts their mental health at risk.[45] The best coping strategy for the first generation is to avoid the culture and language of its adopted homeland altogether. Remaining in constant Facebook contact with the extended family back home might look like a strategy for failure for an immigrant, but it has been shown to protect a newcomer's sense of well-being. Acculturation, assimilation, and other efforts to integrate can expose newcomers to more racism and social exclusion: their language skills are criticized; they are told they are taking jobs from native-born citizens; they have to acclimatize to new cultural practices. All of this threatens mental and physical health. Fortunately, children of immigrants—the second generation—are likely to do much better socially and show fewer signs of stress when they try to fit in with their host culture.

Even on the job, avoidant strategies can work to one's advantage. After all, what do you do when your boss is incompetent? Go over her head and lodge a complaint? Sabotage her efforts to control everyone? Confront her and ask her to change? Or do nothing? This last strategy is common. Employees hunker down and do their work, enjoy time spent with people on their team, and avoid any thought that they can change an unchangeable situation. While avoidance is a poor choice in most situations, when a person is completely checkmated, it may be the only workable option. Is it, though, a sign of resilience? That depends on one's perspective. If an avoidant strategy delivers the peace of mind we need to persevere, then it strikes me as an adequate adaptation in the moment. But truly resilient people know when to change. When a bad boss gets fired or, as is too often the case, promoted, the mark of a resilient individual is his willingness to revert to being the kind of inspired employee he once was.

Another pattern of resilience is what we call *hidden resilience*. A colleague of mine, Dr. Sigrun Juliudottir at the University of Iceland, loves to tell the story of a forty-year-old man whom she remembered from her earliest days as a social worker. The man grew up in an orphanage where he was deprived of food due to shortages and the miserly ways of his housemother. When he was 11, he got was assigned to assist the blacksmith. One day, while the blacksmith was making a second key for the food pantry at the orphanage, the boy managed to make a paper copy when the blacksmith was not looking. He fashioned a replica from wood so that he could sneak food at night and not be discovered. Years later, after locking herself out of her office, Dr. Juliudottir met that same young man. He had grown up and become a successful locksmith, using his delinquent talents to overcome a bad start in life.

Our positive development is the result of our relationships with our environments and how well we are able to exploit opportunities when they come our way, even if that means stealing to survive. Hidden resilience involves adopting strategies that are often overlooked or judged harshly by outsiders.[46] In some circumstances, a young woman's aggression may be a reasonable, albeit unsustainable, reaction to sexism. Likewise, binge drinking among adolescent boys living in extreme poverty may function as a rite of passage when other more socially desirable pathways to adulthood are not available.[47] And forever delaying forming a family may protect some adults from the uneasy feelings of failure, which they are sure await them the moment they try to live up to their parents' expectations.

Finally, resilience can also appear as maladaptive coping. In a family, community, or worksite where access to socially desirable ways of adaptation is blocked, a less attractive but potentially workable alternative can be a bad behavior that brings with it good outcomes. There are more examples of this pattern than one might think. Most, not surprisingly, are controversial if thought of as patterns of resilience. For example, in a neighborhood where there is a real threat of street violence, gang involvement can solve problems of safety and economic insecurity and help marginalized youth resist racial discrimination. Likewise, teenage mothers from very poor communities sometimes experience having a child as a rite of passage that confers on them status as an adult and provides them with a more powerful identity than any other they could create on their own with the resources available.[48] Whether or not a pattern of coping will be accepted as a sign of hidden strength or judged as maladaptive depends on how people in positions of influence judge and label the behavior.

* * *

When we experience a succession of stressful events or a single traumatizing moment, our path to resilience is influenced by our personal strengths, the messages we hear from those around us, and the resources and opportunities at hand. The good news is that most people, when given the choice, choose the more positive and adaptive patterns of resilience. They do not want to be outcasts. Existing on the fringes is unsettling unless one has co-conspirators. These patterns of resilience, however, are not stable qualities: they change over time. To assume that someone who has adapted well once will adapt in the same way in the future ignores the complexity of our misfortunes and the variability of the environments we travel through.

I was thinking about this when I heard about the death of a colleague of mine. He was a retired professor who still came to his office most days to work on volunteer projects. He looked after a few graduate students and was always handy to review a paper or sit on a committee. He was also proud of his Ukrainian heritage. In fact, the day he died, he was with his friends celebrating his national holiday. Sometime before dinner was served, he got up on his game knees and started walking to the kitchen for another drink. He never made it. I heard later that he simply tumbled over and died from a massive coronary. The autopsy revealed that his arteries were badly clogged; he must have known he was at risk of a quick death. I wondered if, on some level, he preferred that to the long, slow descent into obsolescence that was waiting for him post-retirement. Odd, I thought, how someone could be so positive about life in one moment, with so much potential to contribute, yet so pessimistic about aging. There is no one normal path to adult resilience. We are all a strange mix of emotions and strengths.

Each of the patterns of resilience mentioned above shares a similar goal. They all help us access what we need to get by as adults. What each pattern looks like in practice will vary, but what each brings provides more opportunities to experience the 12 resources discussed in Chapter 1. Regardless of how one manifests resilience, we all negotiate with our environments to find the things we need to thrive as best we can, with the resources we have, in ways that matter most to us.

Chapter 3

The Biology
of Resilience

THE FIRST CHAPTER OF Eckhart Tolle's best-selling book *The Power of Now* opens with the story of a beggar sitting on a box. A stranger comes along and asks the beggar what's inside.[49] Implausibly, the beggar who has sat on the box for years has never thought to open it. When finally he does, it is full of gold; thus, the metaphor: we are all beggars seeking something from someone else when everything we need is already there inside of us.

Apart from being insensitive to socially and economically disadvantaged people, and slightly nonsensical, the story of the beggar avoids the fact that much of what we need is not inside us (or under us) but handed to us by our genes, birthright, and social position. Therein lies a challenge for anyone who wants to view resilience as something other than a set of personal qualities that we attribute to rugged (and supposedly enlightened) individuals. How do we do anything about our genes, our birthrights, or our social disadvantage, much less the natural environments that are being despoiled or, in some cases, ravaged by natural disaster? These are hard questions, but research on resilience has shown that even the worst problems are not beyond the control of resourced individuals if we think about changing environments more than changing ourselves.

I learned this from many stories I have heard over the years, although some tales of survival stand out more than others. Akiko had just turned 18 a week before March 11, 2011. That was the day a 9.1-magnitude earthquake shook the seabed off the northeastern coast of Japan at a depth of more than 20,000 meters below the ocean surface. The result was a towering 35-foot wave that destroyed the town of Yamada, where Akiko lived with her parents. She was in a car with a friend returning to school when she heard the tsunami alarm. Stuck in traffic, the car was tossed like a rubber dingy until it became submerged. Akiko's friend was knocked unconscious and drowned. Akiko managed to break the window and swim up through the debris, gasping for air as she dodged floating cars and pieces of concrete falling around her. She was the only one in her immediate family to survive.

A photographer who happened to be on a nearby hillside caught several pictures of Akiko swimming in the very chilly water until she was able to hoist herself onto the roof of a four-story building that had miraculously remained standing. Akiko remembers cutting her hands on the sharp metal of a drainage pipe before she was able to break free of the eddies that kept pulling her under. Once on the roof, she braced herself against a ventilation fan and waited 12 shivering hours for the water to recede and rescue crews to arrive. She recalls being lifted by helicopter through a haze of acrid smoke from the dozens of fires that had started when natural gas pipes ruptured.

By every measure of risk, Akiko should have been traumatized by her experience. When I met her, a year and a half later, she was still having nightmares, but she was attending school, completing her high school credits, and considering her options for postsecondary education. She was housed with an aunt who had lost her

husband. Together they occupied a small temporary home fashioned from portable trailers. Long rows of squat steel-sided units
had been placed end-to-end on a soccer field next to the high school.
Each home had its own hot plate and toilet. Akiko did not have
much good to say about her aunt, but she also knew she did not
have many other options for housing. At least she had been resettled
in her community and back in her high school. That meant she
could spend time with her friends, who shyly spoke of their own
harrowing survival the day that their world had drowned.

A number of nongovernmental organizations had also arrived
in Yamada in the months following the disaster. It was at an NGO
that Akiko and I were introduced. Its program provided evening
and Saturday tutoring for students whose families could no longer
afford to send their children to regular after-school classes. For most
Japanese youth, I was told, the normal school day provides them
with only a small portion of their lessons. Additional instruction is
a part of most children's lives, especially for students whose parents
expect their kids to go on to college. What looked grueling to me
is accepted as a regular part of child development in a culture that
preaches obedience, order, and rote education.

How had Akiko managed to keep going, to avoid the debilitat-
ing, paralyzing effects of living in such a chaotic situation for so
many months? Why had the deaths of her friend and immediate
family, and her own near drowning, not left her with more evident
emotional scars? Patiently listening to Akiko tell her story, I learned
how she and other children and adults like her could be protected
from the more damaging effects of extreme loss and a cascade of
potentially traumatizing events. While Akiko was not exuberant or
deeply insightful, she spoke clearly about the routines she had in her
life and the continuity she experienced between who she was before
the tsunami and who she was 18 months later. There was also the
sameness of her peer group and school environment. Placement

with a family member, though far from wonderful, meant a sense of identity as a member of a family and a culture. The interventions of government agencies and nongovernmental service providers had also given Akiko a sense of hope for the future. No one, it seemed, was providing individual psychotherapy or masking the trauma with sports and other forms of play. Not that these things would have been bad—they just did not seem necessary. As a colleague of mine, Keiji Akiyama, explained to me when I puzzled over Akiko's success, her life still resembled that of many other young Japanese, regardless of the tsunami that had destroyed her town. It was a simple lesson in resilience.

We often hear about victims of disaster recovering best when they are future-oriented, but we conveniently forget that a future orientation is not cognitive gymnastics (an idea constructed alone in our heads). Obliterate the world around us, and one person in a thousand might maintain an optimistic outlook in life, more by chance than design. Put in place the resources needed by an entire population of displaced, traumatized individuals, and the majority will regain normal functioning in a short period of time, so long as those resources are culturally and contextually relevant. No wonder, then, that when I asked Keiji why the service providers were only offering tutoring programs instead of recreation and psychosocial programming, he looked at me confused, cocking his head to the side: "Why would we want our children wasting their time playing?" he asked. It was a shocking revelation. We cannot separate culture from the resources people need to cope with traumatic experiences. As long as Akiko and her peers were being treated like "normal" kids, they were being protected from the dangerous consequences of extreme loss: the physical, emotional, and neurological damage typical in people who fail to cope. A future orientation is nothing more than a realistic appraisal of our opportunities and whether our future looks possible.

★ ★ ★

To understand why our success depends more on our environment than our individual biology, it is worth starting at the very beginning: in the womb. That is where the story of epigenetics begins. It goes without saying that when it comes to genes, some of us won the ovarian lottery. We were born with a privileged set of codes ideally matched to the demands of the place in which we are expected to function. Every culture likes certain phenotypical features more than others (e.g., color of hair, complexion, or body type). But that lottery, we are discovering, is far from fixed. A new wave of epigeneticists is proving that genes are not destiny.

Among those at the forefront of this emerging science is Frances Tylavsky at the University of Tennessee. She has been chronicling how we are changed at both a genetic and neurological level by our experience of our mothers and fathers. One of her studies is using data from a multi-year investigation that has been tracking 1,503 mother–child pairs in Memphis, Missouri, from pregnancy through elementary school. Tylavsky and her team are learning that when we stress mothers, their stress gets under their skin and into the fetus, resulting in predictable changes to children's behavior when they are four and eight years old. Genes, we now know, have switches and are susceptible to environmental triggers that turn them on or off (this is called *methylation*). A mother's stress triggers a reaction on the part of the fetus, which helps it survive, but that same propensity for survival in the womb may compromise the child when these triggered genes change the child's behavior after birth. The most likely outcome is disruptive behavior that even years later prevents the child from succeeding academically. Speculation is that these changes in gene expression are some sort of protective mechanism that the fetus uses to its advantage. Maybe this pattern has, over our evolutionary history, made the fetus better able to cope with the environment it is about to enter. After all, if the mother is experiencing violence, then the chances are good that the child will experience the same after birth.

Tyvalsky's colleagues, including Michael Kobor at the University of British Columbia, have shown that after birth and during the earliest months of a child's life, the frequency with which he or she is held can influence gene expression as well. Leave a child too much on her own, let her feel stressed, and the child's biological development will lag expected milestones. Pick up a child and offer plenty of cuddles, and the child's pattern of development will approach normal once more. These patterns were first noticed in studies of rodents. Mother rats that licked their young produced genetically different pups with greater resilience to stress. Kobor and his colleagues have shown that human babies may not be that different from our mammalian cousins.[50]

What is important about this research is not just that it shows that genes shape behavior. That is an old story. What we are learning is that environments get under our skin and shape the way we respond to stress, changing our thoughts, feelings, and behaviors. Replace a bad environment with a good enough one, and our vulnerabilities can be muted. No motivation is required. Our DNA takes care of us in ways we cannot yet imagine.

While much of the work on genes and resilience has taken place with children, an increasing number of studies are observing similar patterns in the ways that adults and their environments interact. Indeed, even adults can show both genetic and neurological changes related to stress. Mona Fishbane, a family therapist in the United States, has shown that the way couples interact (a potential environmental stressor for sure) changes neurological patterns, making it easier or harder for partners to reconcile differences. As a family therapist changes a couple's pattern of communication, she may be able to influence neurological patterns in chemical release and even gene expression, contributing to better or worse attachments between the spouses. As the theory goes, changes to one system (the couple's communication pattern) affect the sustainability of other co-occurring systems (each partner's autonomic nervous system), which is important to an organism's survival. One caveat, however, is required: these changes take a while. Even after therapy has motivated them to change, couples constantly trigger negative reactions in each other. The wiring beneath each person's reactions is sensitive to the environment, but individuals do not change their wiring as quickly as the words they speak to each other during arguments. Change what is said for long enough, though, and neurological rewiring is likely to occur.

This may make it sound as though stress is bad for our mental and physical health, yet as mentioned in the previous chapter, that is not the case. We are more resilient when we encounter manageable amounts of stress and have the resources required to put our lives back in order. A biological sweet spot (remember the steeling effect?) exists between excessive psychological load and the necessary amount of challenge we need to be inoculated against future calamity. The trick is to make accurate assessments of the stress we are under, avoiding hyperbole about overwork and underappreciation. Experts have shown that not all stress is equal:

- Positive stress is characterized by moderate, short-lived increases in physiological markers like heart rate, blood pressure and the production of stress hormones like cortisol.
- Tolerable stress is a physiological state that has the potential to negatively change brain functioning but can be dealt with if we have supportive relationships, employment, and health care. The stress is real, but so too are the resources we can marshal to move forward with our lives.

- Toxic stress is not manageable. It results in the frequent or pro-longed activation of our body's stress response system. We do not have the resources to deal with it, nor to buffer its impact day to day. Spend enough time trying and failing to manage toxic stress, and we are likely to compromise our autoimmune system, experience a change in gene expression, and become neurologically vulnerable in ways that make us prone to every-thing from addictions to obesity.

The kind of stress we experience will be partly decided by how we label it. Raffael Kalisch from the Johannes Gutenberg University in Germany believes resilience can be boiled down to a simple mat-ter of attribution style.[51] Do we blame ourselves or others for our troubles? Is the attribution correct? Does it empower us or leave us feeling more helpless? It is a tidy explanation, but attribution style cannot fully account for the many complex processes associated with resilience over a lifespan.

A good example of this complexity is the impact of parent-hood on parents. Women joke about "mommy brain," which in fact describes what happens when we are overwhelmed with worry, deprived of sleep, and sensitive to every emotional slight. We become forgetful. We put our car keys in the refrigerator or leave a sleeping child in the back seat of the car. Parents (it happens to men, too) can be forgiven in the fog of dysfunction that comes with having a newborn in their lives.

What is discussed far less is the way newborns offer their adult caregivers an opportunity for neurological and psychological growth. Having a baby changes our capacity to handle stress. We get more efficient at multi-tasking. We expand our emotional land-scape, becoming better at bonding and more motivated to look after others. While not all children evoke the same reaction (that crying baby on the airplane is still a nuisance), brain imaging suggests that in the presence of our own child, the reward circuitry is modified during the postpartum period. Both mothers and fathers become more responsive to their child's coos and gurgles. We also become better at attunement and self-regulation.

All of this occurs because of a series of changes that occur in parents' brain architecture. The amygdala, that most primitive part of our brain, is activated by our child's screams, but the prefrontal cortex that regulates our emotions and facilitates problem-solving also gets a boost in functioning, allowing us to pay attention to our baby's needs and respond with the right amount of empathy. Without this dance of competing functions, parents would not have what it takes to look after a newborn who occupies every waking moment. As with so many dimensions of our adult lives, our capacity to cope with a wide range of challenges is expansive. Add a baby; grow your mind.

For these reasons, it is important to consider the context in which we experience each and every stressor. Context is key to understanding whether stress will enliven us or leave us exhausted. The amount of stress we experience, whether we have the resources to cope with that stress, and to whom we attribute our success afterwards are all key considerations when assessing the likelihood that a particularly stressful period in our life will be perceived as positive, tolerable, or toxic.

Moshe Szyf and Michael Pluess, who have been studying patterns of stress reactivity for more than a decade, point out that when we adapt to stress and cope with the resources at hand, we are likely to feel good about ourselves.[52] This idea is an obvious one, but it is one we need to keep in mind. Stress is good; toxic stress is bad. Find the right balance between stress load and resources, and our capacity for future resilience is enhanced.

* * *

It makes sense, then, to find the most resource-rich environment possible. The odd thing is that we tend to do exactly the opposite of what is good for us. Rather than give our children an environment stockpiled with opportunities to ensure their successful physical and psychological development, we are hell-bent on keeping them away from the experiences they need. There is plenty of good science showing that when children get outdoors and get dirty, their physical and mental well-being improves for a lifetime. As I am writing this, it is a drizzly April day with wonderfully enticing puddles everywhere. I loved walking to school on days like this, black rubber boots slapping against my calves. Each puddle was an opportunity to make the biggest splash possible. The cars that whizzed by doused me and the other kids in gritty cold showers. Little did we know that these antics were beneficial, especially from an immunology point of view. Kids, it seems, need to eat dirt. A well-known article by Thomas McDade and his colleagues at Northwestern University demonstrates that people exposed to lower levels of microbes during their infancy were much more sensitive to the pro-inflammatory effects of stress as adults. In practical terms, the more germs we experience as children, the better our immune systems will handle stress later in life. Dirt makes us stronger and more resilient.[53]

If we think about this as a preventive immunization, giving kids a chance to get dirty may be just what the doctor ordered for long-term resistance to stress. Contrast that advice with the growing number of grocery stores that provide sanitary wipes next to the shopping carts. While I sympathize with parents who are concerned for their children's well-being, I think the microbiologists would say that all that wiping is making our children sickly. The best thing we can do is let our children lick those grocery cart handles (unless you know for certain there has been an outbreak of SARS or a similar deadly disease in your neighborhood).

If children today seem to us vulnerable or unable to look after themselves, we need to stop blaming them and start pointing the finger where it belongs: to caregivers, to alarmist media, to politicians, school authorities, and law enforcers who exaggerate threats and frighten us into believing that the world is hostile to children. Perfectly healthy and loving environments are turned toxic by withholding manageable amounts of risk and responsibility from children. A 2007 British report that traced the roaming privileges of same-aged children over four generations from one family in Sheffield showed that in 1926, eight-year-old George Thomas was permitted to walk 10 kilometers without adult supervision to a fishing hole. In 1950, Jack Hattersley, Thomas's son-in-law, was allowed to walk two kilometers to the woods. By 1979, Hattersley allowed his daughter, Vicki Grant, to wander as far as the local swimming pool by herself, a kilometer away. And in 2007, Grant's boy, Ed, could go only as far as the end of the street, 275 meters from his front door. Do these restrictions imposed on children make them stronger, or does the toxicity of their parents' worry deny children their basic need to take responsibility for themselves and the need for opportunities to build both life skills and physical health?

Our children also need the psychological immunity that comes from an environment that occasionally forces them to self-regulate. Once again, even an individual quality like self-regulation depends much more on demands from our environment than from individual effort. It is an easy pattern to see if one looks for it. On a recent flight back from Florida, I encountered a four-year-old girl coming home from Disney World who fussed for hours because her parents had stowed her headphones in their checked luggage. Her parents started by offering her earbuds provided by the flight attendant; then they tried distracting her with games; and finally, when the screams became louder, they blamed each other for the girl's bad behavior. This unfortunate child had been so indulged that she could not cope with a small inconvenience and amuse herself for a couple of hours. The more her parents tried to make her perfect, the less perfect she was.

To be resilient is not about having our every need met. It is about being given the right amount of stress and the right resources to overcome manageable challenges for healthy biological and neurological development. If those same parents were wise, they would offer their girl opportunities to be uncomfortable more often. I would suggest, rather than a trip to Disney World, a long drive to someplace with low stimulation and no electricity. That might help the little girl learn the self-regulation skills she will need to deal with problems later in life. Change the little girl's environment and she will change her behavior.

To be fair, parenting under any circumstance is a tough job; no manual is provided. But changing environments changes kids (and adults).[54] Here is a personal example from my own years as a parent. When my son, who has now successfully graduated with a degree in engineering, was in grade three, I noticed that he would read out loud what he thought the printed words said rather than what was actually written on the page. Night after night, as I had him read to me before I read to him, I would tell him to read the words. I thought it was his stubbornness, or laziness, that made it difficult for him to do what I was certain he was capable of doing. To make matters far worse, his school assured me that my son was progressing well and that he was simply taking time to grasp a larger vocabulary. At first, that made sense, but three months later, I no longer believed what I had been told and I took my son, at considerable expense, for a special reading assessment. It turns out that he had no word attack skills, or what we commonly call *phonics*. As long as he recognized an entire word, he could read it, but give him a new word that he had yet to memorize, and he simply guessed at what it said. To be fair, he was a great guesser.

When I returned to see my son's teacher and asked her if she could teach my son phonics, she looked at me kindly and explained, "Oh, that's an old way of learning to read. We now use whole language."

"I understand that," I said politely. "It's just that he can't sound out new words and is relying on his memory. I'm not sure he's learning to read with whole language. He is already about a grade level or two behind. Could we supplement his curriculum with something tailored to his way of learning?"

She looked at me condescendingly and told me to be patient; everything would be fine. I instead hired a private tutor to teach my son phonics. Four months later, he was at grade level. Six months later, he was two grade levels ahead. He was bright but handicapped by an arbitrarily rigid system of instruction. Change the system, and soon my son was not only reading better, but he was also enjoying the advantages that reading on his own could give him. With me as his advocate, he recovered. Without me, no amount of coaching was going to make him love reading. I still get a cold shiver down my spine thinking about what his life would be like if he had not been taught phonics.

Unfortunately, so much of popular psychology winds up blaming people for being the victim of problems they might not be able to solve on their own. Pop philosophers like Eckhart Tolle are contextually naive. The fit between our individual strengths and weaknesses and the demands placed on us by our environment are what determines whether we will stumble or regain our footing during a difficult time in our lives. In the language of epigenetics, a *dynamic epigenome* gives us the capacity to adapt to our environment, shaping our thoughts, feelings, and behaviors (our phenotype) in ways that help us to be our best. These patterns of turning genes off and on have one of two effects: they either buffer us from the onslaught of adversity, turning down the volume (so to speak) when we are exposed to risk factors that might harm us, or they give us new energy and more internal resources to cope when all hell breaks loose. Either way, the story of individual biological adaptation to stress is really the story of how much our environment supports or fails us, and the resulting neurological, genetic, and behavioral changes that follow. To grasp resilience, we need to know a person's full story, not just the one inspirational tidbit that affirms the fortitude of the human spirit or a simplified description of the risks that individuals have overcome.

Starting in the 1990s, a group of researchers began studying the long-term effects of ten adverse childhood experiences.[55] Adverse experiences are those which include physical, emotional, and sexual abuse; severe physical or emotional neglect; the incarceration of a parent; a parent's mental illness or addiction; parents' separation or divorce; or domestic violence. As the number of these events increases in a person's childhood, so too does one's odds of developing a host of problems during adulthood, ranging from sexual promiscuity and addictions to drug abuse, heart disease, liver disease, depression, financial stress, obesity, intimate partner violence, sexually transmitted infections, smoking, suicide, unintended pregnancy, sexual violence, and lower academic achievement. To have a childhood experience count toward these bad outcomes, a person only needs to have experienced adversity once before their 18th birthday. The intensity of the experience and its frequency are not considered when assigning an adverse experience score. Whether your father went to jail for a single day or for 20 years, the risk, in theory, is the same.

The adverse experience studies have shown not only that children who are traumatized during their early years grow up to become adults who are at risk of mental and physical health problems but also that the propensity for these bad things to happen will be passed along to their children, resulting in the intergenerational transmission of misery. Sarah Enos Watamura[56], with the Stress Early

Experience and Development Research Center at the University of Denver, considers this the most important public health crisis of our time. Fortunately, it is possible to positively address people's inflated adversity experience scores. Remember when we discovered that lead affected child brain development? Governments at all levels, industry leaders, and health professionals responded, and the problem was solved quickly. Awareness led to solutions: lead was removed from gasoline and paint. The same thing is happening again, as neuro-developmentalists, epigeneticists, and social workers form transdisciplinary networks devoted to preventing child exposure to potentially traumatizing events. Unfortunately, the response this time is slower. Not everyone is convinced of the potential for a healthy return on investment, and our society has yet to dedicate the necessary resources to lower children's exposure to preventable stress, in utero and after birth.

It is frightening to think how predestined our lives can be. It is also enlightening to think how easy it can be to ensure mental and physical health. The only cause for pessimism is when we ask why change has not yet happened on the scale that is needed. Imagine what could be accomplished: Empty jails, fewer people addicted to drugs, less police, the end of the obesity epidemic, far fewer unplanned and teenage pregnancies, better academic achievement, and a healthier, less violent population. These are realistic, achievable outcomes that could be realized in a generation.

The adverse experience studies tell us a lot about risk, but what they tell us about resilience is a little more confusing. First, just because your score is low does not mean you have resilience: problem-free is not fully able. Not experiencing bad things does not mean a person has the capacity to cope when stress attacks. In fact, many people present as vulnerable, because of high adverse experience scores, yet resilient at the same time. They have the social supports and personal management skills to flourish.

Second, a child's vulnerability can change. Adverse experiences occur because the child's environment is flawed. There is nothing intrinsically wrong with the child that causes him or her to be at risk. Change the environment negatively, and the child's odds of engaging in high-risk sexual activity or experiencing depression as an adult increases. In fact, it has been estimated that childhood stress can decrease life expectancy by as much as 20 years, far more than the combined effects of obesity and smoking. On the other hand, positive changes to the environment can quite literally change our inner genome and slow the aging process. [57]

There is another interesting thing to learn from the adverse experience studies. We find that 49% of people who experience four or more adverse childhood experiences develop depression—far higher than the rate for people without adverse childhood experiences.[58] This pattern is the same for sexual promiscuity. Individuals with four or more adverse experiences have a 14% chance of being sexually promiscuous, compared to a 2% chance for those with scores of zero. What is astounding about all these numbers is that the *majority* of people with higher scores never succumb to the risks they face. Most have good lives despite bad starts. How do we explain that?

Robert Anda,[59] formerly of the Centers for Disease Control in Atlanta, was one of the original investigators on the adverse experience studies. As profoundly influential as his work has been, it has not explained why most people avoid problems despite high scores, likely because the science of resilience is still developing, and the question is far more complicated than it seems. It is difficult to know what to measure and which outcomes to track. After all, there exist many different possible outcomes from early exposure to a socially toxic environment (think abuse, neglect, or general chaos in the home). If a large portion of people with a high score does not show one predicted symptom, it may be that a resilience factor prevents that one condition but does not necessarily give the individual immunity from all bad outcomes.

Such problems are what resilience scholars such as myself are teasing apart. Those of us who think about systems have noticed patterns of *multi-finality*. Just because a life course starts in a bad place does not mean it has only one possible outcome. It is our complex interactions with multiple systems that account for our success or failure later in life. Our adverse experience score might correlate with negative outcomes, but we are all immersed in many different systems at once. These other systems can change where we finally end up. A challenging family system might be compensated for by a strong educational system. Likewise, the effects of early exposure to violence may be lessened in a child with higher cognitive functioning. Just because something bad is predicted to occur does not mean that other systems cannot prevent it.

Multi-finality, then, explains some of the unexpectedly good results from people with troubling adverse experience scores—but not entirely. We need to also consider how different levels of exposure to risk influence which resources and interventions are likely to work best for different people. The thinking here echoes what I described at length in the last chapter as the principle of *differential impact*. To fully understand developmental outcomes, we not only have to account for the number of a child's adverse experiences; we also need to know their severity and frequency. This is tricky. In clinical practice, I have met plenty of people who were as traumatized by a single episode of sexual abuse as were others who had been abused for years. In both instances, the abuse is one factor among many that will determine how much trauma an individual will experience. How she thinks about the abuse, the meaning she attaches to it, and the quality of her other experiences in the world will influence how much the experience of abuse affects a person. For this reason, it is important to think about developmental outcomes over time and in context.

As risk increases, it is likely that a person's resilience will suffer. This pattern, sometimes referred to as *inverse collinearity*, is critical to understanding the many different outcomes from early exposure to problems. Think about a child whose parents have divorced, whose father has gone to jail and cannot pay child support, and whose housing and schooling are substandard. As this child's adverse experience score increases, as his problems become severe and chronic, social and institutional supports are also more likely to be fractured, foreclosing his options in life.

While our adverse experience score is certainly not our destiny, it does imperil our future when it puts us in an environment void of the resources we need to buffer ourselves from the bad things around us. Change an individual's access to those resources, and you change the possible outcomes. Once again, rugged individualism has very little to do with it. The good news is that scientists are now working to identify which resilience-promoting resources work best in which circumstances. Unfortunately, no one study has plotted with the same empirical rigor as the adverse experience aspects of a child's early environment that predict a healthy adulthood. Most psychologists, psychiatrists, and clinical social workers would rather study why things break down than explain why they do not. That said, I have been working to isolate the factors that best account for people beating the odds for two decades, as have other researchers around the world. We are getting closer to an answer.

Amanda Sheffield Morris at Oklahoma State University has suggested, for example, that people need 10 protective and compensatory experiences during childhood to be resilient later in life.[60] These include unconditional love from someone, a best friend, opportunities to help others, contact with a trusted adult, involvement in at least one civic organization such as a service club, a religious organization or a non-sport social group, involvement in an organized sport, a school with proper resources and good academic instruction, a clean and safe home to live in with enough to eat, an artistic or intellectual pursuit, and clear and consistent rules. The higher your number of protective and compensatory experiences, the better you can expect to do in life, even if your adverse experience score is already high. Although these experiences begin in childhood, we know from studies of adult development that many of the protective and compensatory factors become available to adults even if they missed them when they were children. Life, we can hope, opens possibilities as we age. We settle in communities; we gain financial stability; we develop new skills. When these good things happen, they can create a cascade of positive experiences and access to resources that ensure our biology is not our destiny. Put succinctly, our lives change for the better when the world around us changes.

Statistical modeling of change processes shows over and over again that individual efforts account for little of the variance between success and failure. The science of success suggests that a visit to the ashram or a full-body makeover is bound to fail unless the change in behavior is sustained through constant effort and reinforcement from a facilitative environment. Without those supports, results are likely to be short-lived or unpredictable. Even after two weeks in the spa, one has to return to the same street, same house, same family, and same job. Epigenetics has shown that when our environments change, remarkable things happen to us at the most fundamental biological level.[61] Our genome changes. Our susceptibility to disease changes. So too does the expression of latent strengths that are lying dormant in our ancestral DNA. It is the richness of our environments that transforms us inside and out.

Chapter 4

The Problems with Positive Thinking

IT IS SAID THAT GALEN, the second-century physician who ministered to Roman emperors, believed that his medical treatments were indisputably effective. Reflecting on his success and failure, he wrote: "All who drink of this treatment recover in a short time, except those whom it does not help, who all die. It is obvious, therefore, that it fails only in incurable cases."[62] This is the way of the self-help witch doctor, telling followers that they are to blame when the guru's advice does not create the expected outcome.

At a conference about resilience in Brisbane, Australia, I shared the stage with a charismatic speaker named Todd Sampson who is a sort of guru to his audiences. An attractive man, he had recently traveled the world training his very ordinary brain to be extraordinary, filming his miraculous acts of courage, endurance, and mind control for a television series. His message was easy and his delivery effortless. The audience of foster children, their families, and the professionals who work with them sat spellbound as he described, among other feats, climbing Mount Everest without an oxygen tank. "Our brains are powerful tools," he told us. "Anyone with a little motivation can train themselves to do great things."

The audience loved Todd, our everyman, in low-rise jeans and a T-shirt. He made neuroscience our best friend and showed us that we could overcome our fears, expand our memory, and commit heroic physical acts like climbing a 120-meter chimney in Utah's Moab Desert, even leaping between two ledges, blindfolded. Only in the final moments of his time on stage did Sampson remind us that these incredible feats needed to be performed cautiously and with the right supports, if anyone dared try them at all.

Another compelling speaker was University of Massachusetts Professor Jon Kabat-Zinn, one of a cluster of Western psychologists and psychiatrists who have been proving that neurological pathways can be rewired through sustained engagement in mindfulness practices.[63] Meditation and other forms of intensive self-reflection and emotional regulation can fix our brains and unlock a tremendous amount of intellectual, emotional, and physical potential.[64] These results can make it seem like the rugged individual is easily programmed if we put in the individual effort. Simple, no?

In fact, the originators of mindfulness training, Tibetan monks and Christian ascetics,[65] understood that their contemplative practices required near complete devotion and isolation from anyone who could disrupt their state of mind. That meant relying on the philanthropic support of their community to clothe, feed, and house the monks while they ascended the spiritual ladder. In other words, even in their bare-bones world, the founders of mindfulness knew that their path to neuroplasticity was facilitated by their privilege as their community's spiritual guides. Certainly, nobody expected the average layperson to achieve the same level of higher consciousness as the monks they honored. It was assumed that as devotees, individuals were too busy raising babies and harvesting the crops to spend their time meditating. The monks pursued enlightenment by devoting themselves to their practice every day for a very long time. That is how meditation can cause neurological transformation that is sufficiently impressive to change our behavior long term.

I am not sure who among us, except for an elite few, has the time and resources to become a bodhisattva, or even mildly enlightened. The woman down the street from me is a divorced mother of two who goes away on three-year retreats where she is perfecting her practice as a Shambhala Buddhist nun. Among the strange things the nuns do is sleep almost upright in coffins to remind themselves of life's impermanence. As you might guess, this woman does not work. Her ex-husband is a successful chartered accountant; I assume his daily grind pays for her path to enlightenment. If I sound cynical, it is hard not to be.

Mindfulness-based interventions (MBI) have become an industry, with businesses advertising different varieties, including mindfulness-based stress reduction and mindfulness-based cognitive therapy. All tend to offer the same four-week or ten-week courses of one- or two-hour sessions, or an online equivalent that involves a combination of meditation practice, group discussion, and the expectation that participants practice what they learn for a few minutes each day. Some evidence shows these techniques work, especially with reducing depression and anxiety in hospital populations experiencing mental distress from physical or psychological health conditions. For example, in one of the best reviews I could find, Simon Goldberg[66] at the University of Wisconsin's Center for Healthy Minds found that mindfulness-based treatments with adults who reported depression, pain, and problems with addictions were just as good as any other treatment, with a documented record of positive outcomes. MBI also outperforms other less well-tested interventions that have inconsistent records of success, and it works much better than leaving people to heal on their own. These results, however, weaken over time. Months after treatment ends, people who received MBI or another evidence-based practice like cognitive behavioral therapy, or a less scientifically sound form of treatment, all report doing about the same as before. Goldberg and his research group suggest that MBI shows a lot of promise as a treatment and is likely as good as anything else we are offering people, but it is not much of an improvement over other kinds of interventions that we already use to help people heal.

At first blush, then, MBI sounds harmless, perhaps useful in some situations. When we look a little deeper at these studies, however, the outcomes are dubious. The studies that have been systematically reviewed almost always include participants who agreed to the treatment and who completed every or almost every phase of intervention. In other words, they were from the start a biased sample of individuals who wanted help. The evidence, on the whole, suggests mediocre results, and even then, only for those who already enjoy the privilege of stable housing, good health care, supportive relationships, and healthy communities. These confounding variables, the factors that make science so messy, are seldom addressed in the papers that get published. When these factors are mentioned, they nearly always undermine the credibility of meditation as an intervention.

Of course, like everyone else in that conference center in Brisbane, I wanted to be inspired by Todd Sampson, but watching his film crew document his astonishing feats, I knew that not one of us in that audience had a hope in hell of climbing a mountain blindfolded or dramatically increasing our memories. It was all a bad misrepresentation of the science of resilience. While Sampson was certainly brave to rock climb blindfolded, he was already an accomplished mountaineer who had reached the summit of Everest before he tried a climb without sight. He was also led up that 120-meter climb by someone he described as one of the best mountaineers in the world, with a crew of technical experts. The camera was on Sampson, while the people who were coaching him were on the periphery. Sampson has more physical courage than I ever will, but to claim he got up that mountain because he rewired his brain is like saying that the airplane I took to Australia got me there on its own; it needed a worldwide network of airports, satellites, government treaties, integrated businesses, and a long list of professionals who spent decades designing and building planes and training to be pilots.

Mindfulness, trauma-informed cognitive behavioral therapy, psychoanalysis, career coaching, and Kripalu yoga are the same. The list of cures is endless these days, and there is big money to be made by insisting people are responsible for their problems. Caught up in thinking about how to change our susceptibility to bad stress, we overlook the fact that our environment is more important than our brain. Few of the studies on mindfulness account for the level of stress exposure experienced by the research participants. It is important to know if patients are depressed, like Woody Allen, living in the lap of luxury in New York, or fleeing civil war. Are they in a country with a 4% unemployment rate or one where 44% of adults are unable to find work? Studies of mindfulness often ignore these inconvenient details.

My team has been looking into what exactly mindfulness practices do for children and adults, and the results are tepid at best. My colleague Raquel Nogueira and I have been reviewing every well-designed study we can find in which a mindfulness intervention was offered to children or adolescents and then evaluated. We are particularly interested in whether young people living with challenges benefit from these interventions as much as children who have normal lives. Most of the studies we have found will not answer this question. Few say much about the risks children face, provide good measures of adversity, or have conducted long-term follow-up. Few have samples larger than one hundred children, which in statistical language means that they are underpowered. Additionally, very few provide detail on the degree to which the children followed through on the homework assignments, which are key to reinforcing the lessons learned. These ambiguities might explain why most of the research on positive thinking, even when found to be effective, has shown that it has a remarkably short period of influence.[67]

One could argue that MBI works best for populations at risk of losing perspective on the seriousness of their illness (as with depression and addictions) and those who are genuinely afraid of death, or who are ruminating endlessly on possible negative outcomes. In situations such as these, common among cancer patients and others with terminal illnesses, time to self-reflect and put life in perspective should be a reasonably useful intervention. In such instances, meditation shows potential to improve lives, though even with a ready and willing group of participants, results remain far from conclusive. A careful search of the literature on how mindfulness training works for adult women who have experienced cancer revealed just three studies that assessed changes in the subjects' levels of depression and anxiety after group training.[68] The problems with these studies were numerous. Some of the anxiety measures that were used showed changes in outcomes, while others showed no change. Some interventions were far more intense than others. The papers never mentioned how the women were selected for each study. Were they rich or poor? Did they have free health care, or were they paying for their sessions? What kinds of supports did they have while they were practicing meditation at home or taking a day away to sit silently? Finally, and this is crucial, were results reported only for those women who completed the training, or do they account for every woman who was invited to attend and those who dropped out?

This final point may seem a bit obsessive, but before we can know if an intervention will actually work in the real world, we need to know how many people actually agreed to participate. To say that "mindfulness-based stress reduction works" and not tell us *for whom* oversells the product tremendously. People who report success with meditation tend to be those who like the technique or are convinced it will work. A huge potential for a placebo effect exists, even when treatment is randomized. Those that complete the training will likely be those with the psychological and social resources to become reflective. Add a charismatic teacher with chiseled abs, and the potential for emotional manipulation increases exponentially. A better study, and one which has yet to be done would contrast mindfulness training with the impact of better access to free health care, a housekeeper, or a steady source of income like disability insurance. In that head-to-head research, my bet is that mindfulness would look weak.

* * *

A better way to think about thinking places less emphasis on selfish goals and more on our relations with other people. Michael Steger of Colorado State University's Center for Meaning and Purpose has found three aspects to meaning in life: significance ("My life is worth it"), coherence ("My life makes sense"), and purpose ("My life has a mission").[69] While each is an individual expression of something profoundly human, each is realized through relationships with others. I experience my life's significance, coherence, and purpose through *both* my thoughts and actions. This suggests that my enlightenment is interwoven with my relationships and responsibilities in my wider community. While there have been historical figures like Nelson Mandela who have managed to sustain their sense of meaning in even the darkest and most soul-crushing places (Mandela was incarcerated for 27 years on Robben Island, a stark piece of rock off the coast of Cape Town, South Africa), few of us have the capacity to survive such atrocities. Most of us will sustain meaning in life by being given opportunities to show others that our lives are worthwhile, to make sense of our experiences, and to reflect on a personal or collective mission. Our lives matter most when we matter to others.

It is practically heresy to question whether or not mindfulness practices like mediation and all the other supposed wizardry that Sampson uses during his stunts create the *sustained* and positive outcomes necessary to change people's lives. Again, when it comes right down to it, only a minuscule portion of what makes us successful depends on our internal resources. Without a never-ending course of treatment and rigorous self-discipline to maintain changes in our capacity to self-regulate, our negative thoughts and feelings can be expected to return.

Changes to our environment are far less likely to disappear over time. People who build a community of concern, connect with a congregation, find a better job, settle their debts, celebrate their culture, and change the way others see them are more likely to produce psychological growth that endures. These changes do not depend solely on personal agency. Once attached to a congregation, for example, one's motivation to attend services can wane for a time without a complete return to social isolation. The congregation will function as a source of resilience, reaching out to its members and reminding them to reconnect. Of course, individual motivation and the exercise of personal agency is important if we are to take advantage of opportunities like someone knocking on our door when our spouse is ill. These opportunities only occur, however, if the social and physical structures that make them happen exist in the first place. After all, only a congregation that accepts new members can be helpful to those new members when they are in crisis. Clearly, it is a two-way street. I need to find spaces and places that accept me, but those spaces and places have to first exist, or I have to create them.

An illustration of this is the emerging science on the impact of divorce on children. The child whose parents' divorce has little power over that decision. At best, he might decide which parent he lives with part of each week. The greatest predictors of successful adaptation, however, are how his parents behave toward one another and the stability of the child's housing, schooling, and community activities.[70] In other words, when adults shape a child's environment to be consistent and stable, children can adapt well to their parents living apart. If there is a history of violence in the home, the child may even experience his parent's separation as a protective factor that increases his lifelong resilience and overall well-being.[71]

A child with the stability I just described who is helped to realize that she played no role in why her parents separated is likely to avoid the self-recriminations that many children experience after their parents divorce. Children need to understand that no amount of wishful thinking is going to repair a broken marriage (though a child's positive thoughts after a divorce can provide a counter-narrative that is more powerful than the "woe is me" thoughts that plague some kids). Happy thoughts make us feel good but will not substantially change our lives unless they occur as part of a matrix of other protective factors that become part of a broader strategy for success.

I know this message resonates with many people and rankles many others who have invested heavily in turning their lives around on their own, but thinking positively or being future-oriented is never enough when we face significant barriers to well-being. I do not blame people for clinging to the myth of rugged individualism and its bedfellow, the cult of positive thinking. And while I have been critical of the naiveté of believing that positive thinking is enough to change our lives, there is no doubt that an optimistic attitude toward life can make us more resilient in some circumstances. On its own, it does very little, but it can be very useful if we have secured other resilience resources.

Take, for example, the inspiring work of Dilek Livaneli, who won the Global Teaching Award for her transformative work in a one-room schoolhouse an hour's commute from the suburbs of Istanbul. A very clever mother of two with more energy than an entire school board, Livaneli showed me first a picture of a rundown building with a bent flag mast and a trampled yard. The photo was from 13 years earlier. To transform that dilapidated school into a beacon of hope for children living on the margins, Livaneli shouted at anyone who would listen that her school needed help, then she cajoled and courted sponsors until she was able to turn her school into a model for what a rural educational facility can be. Children now attend classes in a lovely painted schoolhouse full of 1,001 books (that was their fundraising goal), where parents (mostly mothers) come into the classroom to help with lessons, maintain the building, and prepare nutritious lunches. There are endless activities to engage the entire community, too. An opera company from Istanbul has performed on the front lawn, and every community festival finds a home at the school. Livaneli did not stop there. She told me that she also looked after the mothers, establishing a cottage industry to produce hand-sewn shoes, giving many of these women their first chance to have a bank account separate from their husbands'. Livaneli sees herself as a teacher and everyone in the community as her student.

This kind of success comes from seeing the potential in our surroundings. It comes from taking a positive attitude toward the future rather than being thwarted by the hopelessness of the present. It also comes from a strong enough personality, one that can push back when the community says "No," or "You're a woman," or "Why don't you just teach our children the basics?" But it also needs an environment ready to reward initiative. When Livaneli pushed against the community and got her volunteers, she was upsetting much of the social order. Women who were supposed to stay at home were suddenly at her school, volunteering and then earning their own money. When Livaneli went out and solicited corporate donations for schoolbooks and new desks, she was breaking the mold for public education. But she did it anyway, and the results have consisted of not only a vibrant school for her students and international recognition for Livaneli but also a remarkable change to the wider community.

An experiment like Livaneli's little schoolhouse makes the case that positive thinking in real-world contexts can become a catalyst for political change. The results, however, are messy and unpredictable. Advocates of positive thinking seem to prefer to retreat to controlled experiments to make their point that attitude matters more than opportunity. They love to study undergraduate psychology students and then argue that this motivated, privileged group of individuals can say something meaningful about humanity. Take the work of Amy Cuddy, the Harvard social psychologist who suggested that we need just two minutes of power poses to change our lives.[72] A typical power pose is standing with your arms in the air as if you have just won a race. In laboratory settings, standing like this can alter our levels of cortisol (which is an indicator of our stress level) and testosterone. With these chemicals swishing around, you can expect to feel more competitive at whatever you are doing. Cuddy found that her subjects did better at stress-inducing tasks, such as applying for a job, when they had struck power poses in advance. However, all of the tasks took place in controlled—indeed, contrived—settings. The results proved nothing. In the real world, overconfidence in an interview can backfire, and nothing about the interviewee's situation is perfectly controlled. A little confidence is not a bad thing, but to imply that two minutes of positive thinking is going to increase your chances of getting a job overlooks the many other things that might happen in such a situation.

Yes, where there is real suffering, positive thinking has its merits. The diabetic who refuses to be defined by his disease, who says to himself, "I am a normal person with a disease," is much more likely to maintain good mental health and live longer than the individual who sees his diabetes as a guarantee of impotence, lost limbs, and blindness. We do need to control our wandering minds if we catastrophize every experience we have: "I got into college, but I'm sure I'm going to fail"; "I'm getting married, but it's likely to end in divorce"; "I lost ten pounds, but I'm going to put it back on eventually." A positive attitude toward aging is estimated to have as large an impact on one's life expectancy as not smoking, accounting for an extra 7.6 years for the average American.[73] So we should not throw the baby out with the bathwater, as my grandmother used to say: positivity does have its place. My problem with the pundits of positivity is that they sell the idea that our thoughts are the most important contributor to our resilience when the evidence clearly suggests otherwise. Positive thinking and changing environments need to go hand in hand. Just once, I would like to sit in the audience and hear one of the gurus of mental gymnastics tell us honestly that no amount of positive thinking on its own is going to improve our lives if we are not also fed, housed, respected by our communities, treated fairly, kept safe, and involved in positive relationships that are free of violence. Anything short of that is arrogance.

★ ★ ★

The experience of aging sheds a great deal of light on resilience. With aging comes high rates of self-reported happiness.[74] People over the age of 50 routinely score higher on life satisfaction scales than their children. Positive beliefs about aging certainly help. They are so powerful that they can change our performance on memory tests, improve the physical response of our bodies to stress, and increase the amount we walk when we're older.[75] All of this is in addition to positivity's impact on life expectancy. A positive attitude no doubt helps those of us of advanced age to keep in perspective the ups and downs of life.

But keep in mind that many of us have an abundance of resources by the time we are in our fifties and sixties. We tend to be more financially secure; we have family and community ties; we are often at a place in our careers where we are looking toward retirement rather than starting a new professional path; we are likely to feel a stronger sense of belonging and be more confident sexually. In many ways, the older we get, the more advantages we have.

What if we did not have these advantages? What if our environment throws us a curveball and moves our job offshore or denies us access to health care and the medications we need? In my community, there are seniors' clubs that gather at malls, where the elderly walk and talk and drink coffee. What if one winds up in a place without friends or family, in a community made up of a grid of busy urban streets or, worse, large roads without sidewalks, or country roads where houses are far apart? What if we are struck by a recurring illness? What if one's family supports have all moved away because the one industry town in which they lived was shuttered? Little if any of the research on aging tells us what happens when positive thinking is the only source of resilience.

One of the reasons that healthy attitudes about aging increase longevity is because they make us more open to accepting (and finding) the supports we need to live well. Dilip Jeste, editor of the prestigious journal *International Psychogeriatrics*, has spent a career showing that we live longer if we are immersed in networks of social care.[76] Like chimps in captivity, the quality of care we receive extends life. Put a social animal into a context where sociability is difficult to sustain, as it is for captive elephants, and we will see decreases in lifespan. Without structures in place, we are likely to be "bright-sided" by our own positivity, to use Barbara Ehrenreich's term.[77] In her book of that title, Ehrnereich reminds us that positive thinking alone cannot conquer social disadvantage and toxic environments. Those mega-churches, she warns, will not lift their parishioners out of poverty, but they will make their ministers as wealthy as plutocrats.

* * *

How, then, do we change our environment when the odds are stacked against us? We can remain in a neighborhood where we feel connected and grow our social position in it rather than blindly relocating to a better neighborhood that promises more opportunities but delivers only social and economic exclusion. We can advocate for our child to have access to a better school rather than moving boroughs and in the process destroying our child's network of family and peer supports. These examples highlight a seldom-recognized truth of economic mobility. It comes at a cost for minorities who wind up in better housing but remain socially isolated or, worse, shunned by neighbors who refuse to stop seeing those who migrate up the economic ladder as outsiders.

We have known for at least half a century, since the Stirling County Study[78] in the US, that certain things about our communities make them more likely to prevent mental illness. Socially integrated communities are better for us: they have fewer single-parent households, stronger relationships between neighbors, good leaders, recreational facilities and spaces, less hostility, fewer disasters, lower levels of poverty, and a shared culture. In such a context, a positive outlook on life is in our best interest. It maximizes our access to social capital because opportunities to connect already exist. Healthy communities do not depend on the internal messages people tell themselves, or even on the number of psychotherapists and yoga teachers. These communities are largely a consequence of good governance and progressive taxation, housing, and social welfare policies. If there is a role for personal agency here, it is in whether we as individuals take full advantage of the opportunities such boring aspects of life provide us. In a context made safe by democratic institutions and an accountable police force, it is still up to each citizen to make the effort to know others in her community and to take the time to make a contribution. This is a very old idea. Aristotle understood that virtue was a disposition to act, and that communities rise and fall on the basis of their collective will to look after themselves.

When a forest fire ripped through the town of Lesser Slave Lake in Northern Alberta in 2011, emergency services, the Red Cross, and a small army of volunteers did their best, but more than two-thirds of families lost their homes. The devastation was heart-wrenching. All that remained of entire subdivisions were cement foundations and streets littered with debris, blackened by the heat. People lost everything. It was so traumatic as to seem life-changing for many residents. Interviewed after the fire, residents reported a number of changes in their attitudes toward others.[79] Many committed themselves to achieve important life goals that they had been putting off. Many established new routines and became much more appreciative of the advantages they had in life. Many changed their patterns of communication within their families, softening their approach to parenting, or becoming more loving, engaged spouses. Many looked for opportunities to help others in the community or decided to become less focused on material goods.[80]

None of this should come as a surprise: all of it is the stuff of great sermons and best-selling novels. To be human is to embrace adversity and learn from it. But are these pledges to new behavior sustainable? Will the individual positive mind-sets adopted after a disaster bring success and happiness well into the future? And, more importantly, are they enough to keep people from experiencing post-traumatic stress and depression, or becoming frustrated and angry? The answer is once again mixed. A positive attitude, encouraged by those around us, helps us to heal and cope with the ongoing stress of adjustment. But it has also been found that the single biggest predictor of adjustment after a crisis has nothing to do with prayer, relationships, or a positive attitude. Recovery depends on how quickly insurance adjusters settle claims.

Colleagues of mine who work as social workers discovered that after major flooding destroyed towns at the base of the Rocky Mountains, people who had their claims settled within a year recovered quicker and showed far less stress than those who had to live in hotels and cope with being away from their community for longer periods of time. As a resilience-promoting factor, a quick claims settlement means that people can start rebuilding their homes. It gives them purpose and focus. It rejoins them with their communities and gives their children the chance to return to their schools. It also decreases the daily stress of living and the worry associated with an uncertain financial future.

The banks and insurance companies must have taken note. When wildfires destroyed 2,400 buildings and forced the evacuation of 88,000 people in another Northern Alberta town, Fort McMurray, in the summer of 2016, residents were scattered across nearby towns and cities and packed into community centers. Financial institutions loaded their staff onto large buses, the kind that touring rock bands use, and on each bus were bank machines, loans officers, and insurance adjusters. The bankers traveled to the shelters, sleeping on the buses so they would not burden the scarce local resources. By traveling to people who had been forcibly displaced, the bankers were able to give their customers access to cash and an opportunity to start the paperwork required to submit a claim for compensation. The effort must have expedited payouts, because people were back in Fort McMurray and rebuilding within months. Not everyone was fortunate enough to have his insurance paid out quickly, but for those who had a friendly banker and an insurance adjuster make a visit, emotional outcomes were likely better than expected. After a major disaster, the first responders should be the fire department and paramedics. Second should be insurance adjusters and bankers. A distant third should be psychologists, and only if financial claims cannot be settled quickly. Mental health professionals, like me, are sometimes needed—just not as much as we think.

★ ★ ★

The worst aspect of positive thinking is the degree to which it promotes and masquerades for victim blaming. It implies that resilience is the responsibility of individuals alone, a simple ideology dressed up to look like science. I learned this from one of my trainees, a woman in her late twenties named Justine. I had asked her how she became interested in the topic of resilience. "Well, it wasn't any one thing," she said. "But a big part of it was when I worked as a relief worker at a children's shelter. Most of these kids were there because they had really crappy parents. I was from white-bread-and-bed-by-eight-on-school-nights suburbia. It was shocking. Really shocking that kids were raised like that. They'd tell me about drug paraphernalia on the kitchen table and all I could think about was my mom tormenting my father to get his 'Goddamn Saturday paper off the kitchen table so we could all sit down for brunch.'"

"Well, anyway," Justine continued, "they eventually assigned me to double staff these three fifteen-year-old girls who kept running from the shelter. They got put up in a motel so their beds could be given to kids who showed more interest in participating in the programming. So there I was with three girls in a Motel 6 next to a downtown strip mall, and I'm supposed to keep these girls in school and, according to my supervisor, 'Appropriately engaged in activities that will further their life goals.' Great, I thought. Like what the hell does that mean? So I'd get them to climb on one of their beds and I'd ask the girls to close their eyes and follow their breath. When they just giggled and made grunting sounds like they were reaching orgasm, I'd bring out drawing books and huge sets of colored pencils and ask them to imagine their futures. I kept reminding them they could draw anything they liked and that I'd help them build ladders to reach the stars."

"And what did they draw?" I asked.

"Mostly they liked the black and blue pencils to draw corpses or red pencils for menacing eyes. They drew their social workers ripping their clients' arms off, or pictures of zombies eating their teachers. But I kept trying and trying and then trying some more, until it occurred to me that they had no pictures in their heads for what was possible. It was like they were climbing a ladder and were on the first rung and the horizon was very near. I had it all backwards. I needed to get them up the ladder first so they could see a landscape of possibilities. I had started inside their heads. I should have begun by opening up possibilities around them and letting them decide which to choose."

"So what did you do next?" I asked.

"Nothing, really. I had zero budget to do anything with the girls and there were all these rules about leaving the hotel. It took a monstrous pile of paperwork to go for a walk. So we watched TV and I tried to inspire them with reruns of Ellen. And then one day, one of the girls, Gabriella was her name, didn't come back after school. My bosses told me to just wait before filing the paperwork. AWOL teens in care are pretty common but the department hates telling anyone how common. But by then, it didn't matter because some municipal garbage collectors had found some of her clothing under a tree by the river a few hundred yards from the motel and called the police. When they sent in divers, they found her, tied and gagged with the rapist's t-shirt. I was told that before she died someone had put out a cigarette on her breasts, maybe more than once. And that was pretty much the end of that career path for me. I came home, cried a lot, ate potato chips and watched entire seasons of *The Big Bang Theory*."

"That's an awful story," I said, meaning every word. I knew this was the kind of experience we all have some time in our career if we are devoted to helping those on the fringes of our communities. We talked a bit more and I reassured Justine that she was still doing good work, even if it was not at shelters. She nodded and then distracted herself by fiddling with the pen she was holding. Finally, I asked, "No revenge fantasies? No grandiose visions to change the system?"

Justine flushed red. She had experienced those thoughts for months.

Storm the barricades? Not likely . . . I'm not the rebellious type. At least not the slash-and-burn, Occupy Movement type. I'm not even the girl who sits on the street holding a flower and begs the police to stop firing water cannons. I never even told my supervisors what I thought killed that little girl. But it did really bug me that the tools I had to help those kids were so pathetically inappropriate. They needed a grandparent to give them a hug. And tutors. And they sure as hell shouldn't have been left to rot in a motel room with some inexperienced counselor as their jailor.

"But this experience made you interested in resilience?" I asked.

"Resilience offers hope," she said.

"And what makes us resilient?" I asked.

"If I had to name one thing, I'd say control. Those girls had no say over what happened to them. That's why they ran away, used drugs, had sex. At least doing those things, they could control their bodies."

It was exactly what I had been finding for years, through both my clinical work and research. Experiences of control make us resilient; and control is something others give us or deny us. We can survive for a time without control, much as prisoners do in jail, but we seldom thrive when we feel disempowered. Control, or efficacy, if one prefers the psychological term, is about experience and perception. We all do better when we experience it in healthy doses. Our well-being improves when we decide on the order of tasks we have to accomplish at work (filing, returning messages, and attending meetings). At home, it is our choice whether we build a man cave or she shed. We want to choose our furnishings and decorate for the holidays whatever way we like. We want to decide whose in-laws we visit at Thanksgiving. We want to have some say over which school our children attend. A crazed hockey parent will time the number of seconds her child is on the ice just to feel in control.

We also do not like being told how we feel or what to feel. Damn those therapists who want to extinguish all anger or anxiety! I know many a person who would rather be fidgety and creative than medicated or mindful. The famous psychiatrist, Elizabeth Kubler Ross, once said, "I'm not okay, you're not okay, and that's okay." Part of being in control is challenging others who tell us how we should feel.

Control is vital to our capacity to overcome adversity.[81] An unappreciative employer, a bad economy, an abusive spouse, an unfair teacher can all take control away from us, and in the process undermine our capacity to cope when life turns sour. So we look for control. We do whatever we can, within the boundaries of our responsibilities, to feel personal efficacy. Control is one of those big hairy beasts that inhabit our lives, constantly thumping its chest for attention. Of course, control must negotiate with structure. The two are opposite ends of a teeter-totter. No one is going to let his three-year-old choose his own bedtime, or allow his father-in-law to come over any time he wants and turn on the television. We insist on rules and reasonable behavior to ensure that an individual's need for control does not destroy those around us. (The behavior of three-year-olds and fathers-in-law can, after all, get out of hand.)

People who feel in control of their lives experience less stress and are more productive. The science on this is clear: attributions of self-efficacy are key to feeling empowered and healthy.[82] If I attribute my success to myself, or my group, I'm much more likely to cope with disappointments. Unfortunately, we tend to think this experience of control is entirely within us. We love to believe in ourselves—in our individual capacity to succeed.

Our minds are wonderfully competent at imagining ourselves as cleverer than we really are. We hate to think of ourselves as victims of chance and circumstances beyond our control. Most of the control we enjoy, however, is that which the world, in the form of our parents, our bosses, and our governments, permits us. Control is always limited or enhanced by the resources available in our environment. It is always circumscribed by the harsh realities of our daily existence, regardless of how devoted we are to positive thinking. I love to think I am in control of my life. I love to think of that life as being full of adventure and fun, hope and change. But I also know those thoughts are easier to sustain when I am safe and loved.

* * *

Years ago, I read an odd little study that reported that crack-addicted mothers who get pregnant tend to be more successful at beating their habit and remaining clean than young women who are not pregnant. There is something powerful about the change in their perceptions of themselves from youth to mother that makes it much easier for a young woman addicted to hard drugs to modify her behavior. I have heard a variation of this same story from cigarette smokers who gave up the habit after their children were born. One mother told me it struck her that something had to change the day she sat with her one-year-old in a doctor's office and wondered who was smoking when the signs clearly said it was not allowed.

She sniffed the air a little deeper and realized that her baby's wrap reeked of smoke. She gave up the habit that day.

We can do great things when our thoughts guide our actions, and we can change our thoughts when the world around us forces us to change. What would have happened if smoking in public spaces was still permitted? The right amount of government intervention can help us realize our best selves or prevent burdensome levels of disease and disorder. When it comes to helping people self-regulate their cigarette smoking, grotesque pictures on cigarette packages, heavy taxes, and "No smoking" signs everywhere are more effective at changing behavior than encouraging people to kick the habit on their own. Expose smokers to an environment that makes their behavior socially unacceptable, and motivation to change rises dramatically. My guess is that without these social inducements to quit, that mother of the one-year-old would have continued smoking, and her child would have been that much more likely to suffer the consequences.

There are times, of course, when we cannot change the world around us, and the only thing left for us to do is change our thinking. Elie Wiesel's reflection on his imprisonment in the Birkenau, Auschwitz, and Buna concentration camps from 1944 to 1945 is a testament to the power of the mind and its capacity to maintain hope when we are at our deepest level of despair.[83] In such instances, changing our thinking is the only choice we have. Whether it keeps us alive or simply makes our passing less painful, there is no way to be sure. What we do know is that through reflection, prayer, and the repetition of devotional practices, we experience our lives as being under our control, even when they are not. Changing our thinking, however, can only ever be a good strategy in a crisis if nothing else is within our power.

So go ahead: meditate; practice Qigong; buy a yoga mat; cleanse; eat more kale; or dance your way through your day. Don't worry, be happy. I have been enamored with all of these strategies (except the kale) at one time or another, but the truth is that the real sources of change in my life have always been found in the world that surrounds me and my ability to marshal these resources when I need them.

Chapter 5

You Are Who You Know

A N ENTIRE FIELD OF STUDY exists that looks at people who experience exceptional psychological, physical, and social growth in contexts where almost everyone else fails. The most famous example is a study of Vietnamese families in the early 1990s living in villages where children were known to be severely malnourished. Only a few families managed to raise healthy children with the meager resources available. Were those children more resilient? Biologically different? Neurologically unique? Or were the families better functioning and the parents smarter than their neighbors? No simple explanation for the differences between the children could explain why some did better than others, but the more researchers dug, the more they could see that the children who were adequately nourished had parents who fed them foods like sweet potato tops and small shrimps that grew in the rice paddies. All these foods were freely available to anyone who wanted to harvest them, although few families made the effort. Successful families also fed their children four small meals each day rather than the traditional two larger meals. Additionally, these families were physically closer to their children. Even the very poorest of families doing these things raised healthy children who outperformed their peers, including peers from households with higher incomes. In part, their positive deviance—meaning that the children deviated from the norm in positive ways—came down to beliefs and opportunities. Successful families had beliefs about what children needed, and they took advantage of what the environment offered. Families with other beliefs, who ignored the abundant resources within their reach, raised undernourished children.

To this point, we have discussed how our individual genes, neurobiology, thoughts, and habits affect our resilience. The closest within our environments—our families, friends, and colleagues at work—can also have an enormous effect on our collective capacity to thrive. Change the functioning of the family, peer group, or work team, and individuals are more likely to show resilience, even if their larger world is seeming to become more volatile, uncertain, complex, and ambiguous. I would argue that our world today is actually more secure and less violent than it was generations ago. We have social insurance programs, crime is down, accident rates are down, there is relatively less war, and there is no need to hide inside our homes to keep safe. But people *feel* more vulnerable and uncertain, which is why our relationships matter: they shape our experience of the world.

In my experience, if I were to ask 10 people on the street what makes us successful or resilient, six will say "perseverance," "motivation," or "faith." In a word, *individualism*. And the last four? They'll say we need relationships. Family, mostly, or, more specifically, our mothers and lovers. Intimate relationships of the kind we have with our parents and later our lovers are extremely important. We privilege them with a magical aura because in a crisis, people with emotionally sustaining connections appear more likely to thrive. Loving relationships are the basis for a disproportionately large amount of psychological research. For the most part, these studies confirm what is intuitively obvious. Close connections with others can help us succeed when they meet our psychological, social, and economic needs.[84] They are among the most important factors contributing to healthy child and adult development.

Many people will remember the Romanian orphan children who were left languishing in their cribs during the 1980s.[85] Both their neurological and physical development was stunted by months or years of neglect. Once adopted by families, especially families with the capacity to care for them and form emotional attachments, the children's developmental trajectories changed dramatically. They moved from being developmentally delayed to functioning at the same level as other adopted children.

Feeling loved helps us to endure pain and to grow emotionally, spiritually, and physically. But these close relationships also give us a very practical edge in life. They keep us fed and healthy; in the case of the miners I mentioned in Chapter 1, or the Vietnamese children above, our families help to keep us alive. No matter how old we are, intimate relationships help us better withstand life's stressors and succeed where others are likely to fail.

A famous study by two Harvard psychologists, John Laub and Robert Sampson, makes this point as well.[86] Laub and Sampson felt like detectives when they found in the basement of one of Harvard's buildings files for a study started in 1939 by the husband and wife team Sheldon and Eleanor Glueck. The Gluecks had identified five hundred boys housed in a Boston reformatory school and followed their development for 25 years to figure out what made some children succeed while others continue their delinquent behaviors into adulthood. Laub and Sampson picked up the study decades after it had ended. By that time, those young men were in their late sixties and early seventies.

As might be expected, many of the boys did reasonably well in life once they hit their adult years. Things that had brought them trouble in adolescence, such as having sex or drinking alcohol, were legal in adulthood. They made a relatively easy transition from juvenile delinquency to being hardscrabble members of their communities and stayed on the right side of the law. Of course, other young men continued to have problems as they aged, and this is where the story gets interesting. There were many things that made a difference to the life trajectories of the boys that did well. Among the most noteworthy was military service, which provided structure and a position of status in society, and new peer relationships and a powerful identity, all things critical to resilience. But these young men also did better if they had "married well." In other words, their success was predicted by whether they had managed to find a supportive spouse who could help them overcome a rough start in life.

I am guessing no one spoke to those men's wives about their experience of dealing with a bad-boy husband. Nevertheless, the point is that a spouse can substitute for a parent and improve the odds that a neglected or delinquent individual settles down and accepts adult responsibilities. It is important to note that marrying well was just one reason why these men's lives turned out well. They used the stability of their relationships to hold down jobs, maintain a place in their community, and, in many other ways, build supportive networks. What if their marital relationships had become strained? Would that have meant the men would go back to their errant ways? This outcome is unlikely if they had made it to midlife without serious psychological impairments and had managed to build a network of social supports.

Studies such as this one prove that intimacy with others skews our developmental trajectories, hopefully in a good direction. Amazingly, our intimate relationships can even change something as individual as our sexual desire. Forget hormone and pheromone levels: a happy sex life has more to do with the quality of the intimacy we experience with our partners than anything inside us. According to family therapists like Pat Love, couples spend as little as 35 minutes a day communicating with each other. The intimacy of those few minutes determines how much sex the couple experiences.[87] Results vary, but most studies show a bimodal, or two-hump (excuse the pun), pattern, with couples reporting sex either twice a month or twice a week. Oddly, the more couples share a lifestyle and the more intimate conversations they have, the less sex they report. It may be that couples these days are becoming more like friends than lovers, trading passion for familiarity.

Explaining the relationship between success and intimate relationships can draw us into the same trap laid by hawkers of rugged individualism. Just as individual motivation is never enough to account for success in toxic environments, intimate relationships provide an insufficient explanation for why we succeed when psychological, social, and economic resources are in short supply. Our resilience is far too complex to depend on feeling loved. We can focus too much on attachment as the source of our happiness. Relationships do not need to be intimate: a dear friend, or a close connection with a therapist, can offer a healthy substitute. We can build resilience out of many other resources that sustain well-being, such as a sense of one's cultural heritage, community, personal identity, experiences of power and control, safety, and social justice. When it comes right down to it, intimate relationships are not one-size-fits-all, and lack of a strong intimate relationship need not forecast a lousy life.

To some scholars, even hinting that intimate relationships are not the key to resilience is breaking an unspoken code. To suggest that mothers, fathers, wives, or husbands are not the single most important external factor to maintaining our mental health threatens the foundations of Western psychology. Yet, if you look for them, you will find vastly different family forms all around the world.[88] I was inspired by a talk I once heard by the Greek cross-cultural

psychologist James Georgas. He suggested that our morbid fascina-
tion with the negative outcomes associated with single parenthood
hides the real reasons for these families' troubles. Step away from
our Eurocentric bias toward nuclear family values, and we will see
that the real factors that put single-parent families at risk have to
do with financial instability and the frequent lack of social supports
that single parents experience. Georgas suggests that our cate-
gorical minds quickly slot single-parent families into the "at-risk"
category based on relationship status. This is only true if we fail
to ask another, more important, question: "If tonight at midnight
you had a crisis and needed help, how far away does someone live
who could support you and your children?" Such a person might be
a neighbor across the street, a family member in the unit upstairs, a
friend renting a room, a professional care provider, or a lover.

Being a single parent is not a risk factor in itself. The cascade of
disadvantages that follow are the real problems. One can argue that
a single parent is in a better position to raise kids than a married
spouse who is exposed to intimate partner violence. Plenty of adults
tell me that when they were growing up and all hell broke loose at
home, they turned to their teachers, coaches, grandparents, or even
friends' parents for emotional and instrumental support.

There are entire societies of psychologists who will fight me
to the bitter end on this point. They will insist that our resilience,
or at least our psychological health, depends entirely on our connection to a committed, caring other. But under stress—and I am
talking big stress, like job loss, domestic violence, or a long drawn-
out illness—people do better when they have many different types
of relationships and many different kinds of resources.[89] In fact,
the very people we are supposed to rely upon during a crisis (our
mothers, fathers, wives, and husbands) are usually completely flum-
moxed by their own emotional turmoil when the people they are
supposed to be looking out for are in jeopardy. To expect these

intimate others to also be responsible for our mental well-being when they can barely maintain their own is unreasonable.

* * *

When the war in Kosovo ended in 1999, a group of American mental health professionals heard that the country was struggling to cope with large numbers of people with severe mental illness caused by the genocide they had witnessed. Few mental health supports were available in Kosovo, and certainly nowhere near enough were present to help the many patients needing care. The worst stories were of families who had to chain their loved ones to their beds to keep them from wandering away or committing acts of violence.

Stevan Weine, of the University of Illinois Center for Global Health in Chicago, traveled with his colleagues to Kosovo, talked to families about their experiences as caregivers, and worked with health care providers to develop solutions.[90] His team invited families of patients to gather in multiple-family therapy groups. These groups offered mutual support and people in them shared strategies for dealing with their loved ones. Whenever possible, professionals worked with these groups and made home visits to put creative ideas into practice. This proved simpler and more practical than offering individual therapy for months on end. The intervention worked because it bolstered the natural capacity of families to look after their loved ones while giving them the structure and occasional access to skilled professionals to solve more complex problems. Of course, there was a risk of imposing solutions to mental health challenges that were better suited to Chicago than Pristina, so Weine and his colleagues were careful to show sensitivity to the traditional extended-family focus of Kosovar society. Local professionals, too, benefited from the training, eventually inheriting the program and expanding it widely. Of the families participating in the original intervention, 93% reported positive results—a number that dwarfs similar efforts even in countries without recent histories of genocide.

This example reminds us that we benefit from a variety of relationships. Our bias for a nuclear family structure is what leads us down the intellectual rabbit hole of attachment theory. I am sure that if Freud had been an Indigenous Elder from a remote Native American tribe or a Chinese patriarch in a large extended family, he would have laughed at his small-minded theories that supposedly explained our relationships. Abundant proof states that we need many different relationships to cope when our lives become stressed—after a divorce, after our partner dies, after we move out of our parents' home for the first time.[91] This is one of the many secrets to resilience: our success depends on us building a network of supports and keeping that network as well tended as the gardens of Versailles. Sadly, most of us neglect our responsibilities as gardeners. We think of our work lives or home lives as narrow spaces where we compete for limited resources. We impose a myth of scarcity when, in fact, we are far more intertwined in social networks than we recognize.

I love seeing this interdependence in action, especially when the assumption is that people are acting autonomously. Take, for example, the sprawling and historic Grand Bazaar of Istanbul. While there, I had the illusion that I was negotiating with each merchant independently as I searched for a gold and diamond bracelet for my partner. I thought I had found a pretty good buy that would impress her immensely and was told by the shop's owner that it was made by the craftsmen working exclusively for him. It all sounded beautifully exotic until I went to another shop to compare prices. There, at the second shop, I asked for a similar gold and diamond bracelet that was not on display. "Oh, yes, we have one of those, but it is in our vault," the owner said, and quickly dispatched one of his shop clerks to get it. A moment later, the clerk returned with the exact same piece that I had seen in the first store, available at a slightly higher price and once again offered with the reassurance of its unique craftsmanship. Ask the store owners about this apparent duplicity, and they will say that both stores are owned by members of the same family, but in truth the network is more complex. The bazaar is a resilient structure for commerce because it avoids outright competition between merchants, none of whom can afford a full stock of high-priced items. In this context of limited resources, they strike strategic alliances and share resources. Competition still exists, but primarily on price.

We live much of our lives in our communities, however we define those. Our lives are not merely family trees. They are networks, ever-evolving, fluid, and, best of all, limitless in their potential. Even if that potential is realized online rather than face-to-face, our relationships are the source of our identities—identities that we want to perform for others.

While working in Moscow with a colleague of mine, Alexander Makhnach at the Russian Academy of Sciences, I spent an evening at the Bolshoi Ballet. It is a grand old theater not far from the Kremlin in the heart of city. Throughout its history, it has hosted some of the greatest dancers to wear pointes. The gilded accents and hand-stitched costumes transport you back in time. Even if ballet is not your thing, it is hard not to be impressed by the athleticism and grace of the performers.

What struck me most that evening was how often the dancers interrupted their performances to take bows. Each time one of the company leads did something outstanding, the music stopped and the dancer swooped down to the front of the stage for a swanlike curtsy while the audience roared its approval. After a suitable period of adoration, the orchestra played, and the ballet continued. It is remarkable, I thought, that such artistry receives such appreciation.

Only, it did not end there. At the end of the ballet was the grand curtain call. A massive red velvet cloth fell and rose again to reveal the company of dancers perfectly placed to accept its standing ovation. The curtain went down and up again, revealing the dancers bowing humbly once more. More flowers were thrown, and the applause continued. After the third curtain call, I began to look for the exit, except the audience did not move. The curtain rose again for a fourth curtain call. Then a fifth, sixth, and seventh. The audience now began to thin, but there ar least two thousand people were still clapping. In all, there were 12 curtain calls, and each time, the company graciously accepted the praise.

Afterwards, on the cold, dimly lit streets of Moscow, I thought to myself how remarkable that had been. I marveled at the confidence of the artists who demanded recognition from their audience over and over again. The next day, I commented on what I had seen to my colleague, and he rolled his eyes. "Yes, yes," he said, "they think they are royalty."

I tell my colleagues that health care providers should act like Russian dancers. The next time a social worker, physiotherapist, or nurse performs a wonderful intervention, changes a life, or brings someone through a family crisis, she should stand in her reception area and announce, "I just transformed a life!" and curtsy, "Thank you, thank you!"

I say this in jest, but being our best selves and preventing burnout requires that our work is recognized for the value it provides. This is true whether we are mental health professionals, bus drivers, or dancers. It is the rare person who seeks only internal gratification. Whether recognition comes in the form of a memo, a friendly chat, or an annual bonus, it makes people feel like their effort is valued by others.

★ ★ ★

Identities are both the stories we tell about ourselves and the stories others tell about us, which join together in our minds to create impressions of who we are.[92] In other words, they, too, are shaped by our environments. Our workplaces, in particular, are crucial to our identities. Many of us identify who we are by the work we do. It is this pattern of identity formation that makes retirement so stressful. Not only do retirees lose their sense of being part of a tribe, but they also lose their sense of control, their routines, and often the meaningful relationships that made them feel that their lives had purpose. Without healthy substitutes, retirees are at risk of dropping dead shortly after leaving full-time employment. The Walmart greeter and the Kiwanis Club member both show post-retirement success stories. These people have found substitute places to belong, feel useful, and remain visible members of their communities.

Psychology has taken a narrow view of identity and tried to convince us that it is an internal experience of the self, one that we can manipulate on our own. In contrast, the sociologist understands identity as something we build together. We know ourselves through our interactions with others, selecting from a limited number of self-descriptions the most powerful one we can find. Do you ever wonder why children all want to grow up and become the same thing as their parents, their teachers, or maybe firefighters, doctors, and nurses? These are the identities most readily available to them. As children grow, their choices broaden, especially if alternatives are perceived as powerful. Biotechnician? Mechanical engineer? Journalist? Researcher? Underwater welder? Entrepreneur? These are work-related identities that children discover as they participate in wider and wider conversations about what people do and the status each job has. For some, it is not the job description that brings status, but the working conditions. I have heard people distinguish themselves, as office workers, from their peers working in factories or on farms. Even if the money for working with one's hands is better, the office worker enjoys a higher status working with her head. Some might say that the difference in a job's status can be determined by whether one showers before or after work. Of course, all of this depends on who is part of the conversation and what they value.

No single identity is necessarily better than another. People who show resilience build powerful identities and find social networks that affirm those identities.[93] When bad things happen, these identities become part of the glue that helps us hold our lives together. They also draw us into relationships. Identities can bring with them the security of routines and consequences as we do what our identity demands we do.

I cringe when I attend workshops and the speaker has the audience recite affirmations: "Every day in every way, I'm getting better and better." It is so untrue. We only get better if everyone around us sees us as better and tells us so. Have you ever thought you were having an okay day, but then a friend tells you that you look tired or stressed? Maybe your complexion is off, or you did not wash your hair. Instead of feeling fine, you wonder why your friend would say such a thing—and you begin to doubt yourself: "Am I really okay? Did I get enough sleep?" Our identities pivot in the wind of social approval. The identity we think we have is never ours to hold or mold alone.

Most people who go through a divorce experience this unpleasant disruption of identity. There is the loss of the successful marriage and what it symbolizes: one may no longer be able to say he or she is a stable, loving person. Additionally, there is always the suspicion that the dissolution of the relationship must be the result of personal failings. Self-doubts become worse when friends take sides: when one spouse is identified as the victim, the other must have done something horrible. If the story about why the relationship ended changes—"oh, *she* had a lover on the side"—identities also change. Identities are fluid, subject to perceptions.

Every one of us, from early in our lives through our senior years, at work and at home, builds and sheds identities. This is a natural process of adult development that makes us more resilient as our environment changes. If one is fortunate enough to live in an environment without disruptions, then it is possible to hold one identity for a long time. But in our postindustrial age, with its shifting cultural landscape, it is difficult to remain singular when there are so many opportunities to become someone else. Hairstyles change; clothes change; work opportunities change; values change. Suddenly our children no longer admire our vintage Camaro; they criticize us for living in our large house in the suburbs; greenhouse gas and energy conservation, they tell us, are the new priorities. With each new attitude and behavior, we unintentionally court a different perception of who we are. Whether that identity is retro and cool or environmentally irresponsible depends where you stand and the feedback you receive.

A tattoo is a wonderful example of an identity being performed. In my role as family therapist, parents often ask me, "At what age should I let my child get a tattoo?" Assuming the child respects her parents enough to accept some limits on her behavior, my advice is usually to tell the child she can have a tattoo when she wears her favorite piece of clothing for more than a year. Like all of us, we build our identities through external cues. A tattoo tells people just as much about us as where we work or what dish we choose from a menu. Our choices position us higher or lower in relation to others. A tattoo can increase a child's status, but only temporarily if the meaning of the tattoo becomes passé.

A successful identity offers us stability when our world threatens to fall apart. It becomes our North Star, reminding us what we want out of life. Donald Trump may be many things to many people, but he has always exhibited a core identity as the flamboyant entrepreneur, through all the ups and downs in his fortune. My neighbor who lost his job in his late fifties also has a secure identity. Even as he adjusted to becoming a househusband and piecing together small contracts to make a living, he never lost his standing in our community as the good-hearted Irishman whose St. Patrick's Day parties are legendary, and whose backyard fire-pit is the place to gather on clear summer nights.

If we are fortunate, like my neighbor, we can have more than one powerful identity at a time. When one fails, the others become more valuable. It helps that the older we get, the more likely we are to find a peer group that appreciates what is special about us. As the sexual advice columnist Dan Savage reminds adolescents who feel suicidal because of their gender nonconforming sexual orientation, "it gets better." To be resilient is to find a place where we can be ourselves and be appreciated for the contributions we make. For the dancer in the Bolshoi Ballet, that may be easy; for the rest of us, it takes work.

When a research team I led studied young people during the decade after high school, we found a most unexpected result. Youth who set their course early in life, those we called navigators, were seldom happy in their late twenties.[94] These were every parent's dream children, kids who had decided early what they were going to be and stuck to their plan. Then went to school and, as expected, became doctors, engineers, and accountants. Some became welders or businesspeople. They should have been happy, but they were not because they had foreclosed too early. Can we make an informed choice during our teen years about whom we want to be as an adult? It is hubris to think so. All we can do is choose our first adult identity (become *something*), then wait until another identity presents itself.

During that study, I met a young man with a wonderful artistic ability who could never bring himself to become a full-time artist.

His parents, first-generation immigrants, had warned him that he could not support himself with his talent, and he had swallowed their logic. He passed his time at college uneventfully, graduating with a useless bachelor's degree in a discipline in which he had little interest. He was lost and he knew it. He might have wound up terribly depressed if he had not come across a strange job ad online. A company was looking for someone with artistic talent, promising good pay. To the young man, finding both an artistic outlet and a paycheck in one job seemed like nirvana. The work? Designing implants—teeth, that is. Who knew, but every tooth has to be individually carved and colored to perfectly match a patient's mouth. Not only is that young man now an artist with a job; he also has the respect of his parents, who tell their friends he works in dentistry.

We are always somewhat at sea with our identities. They are complex amalgams of different parts of our lives. We do better when we have opportunities to show others there is something special about us. Sometimes, that sense of being special comes from work; sometimes work is nothing more than a steady income. In the latter case, we do best when we follow our passions for travel, sports, cooking, or something else on evenings and weekends. Some of us will even turn these passions into employment. The important thing is that our identities are strong and flexible enough to help us weather the inevitable storms.

Not all identities are socially desirable. The jihadist, for example, is anxious and insecure about his identity. Violent radicals need others to notice them.[95] So, too, do narcissistic bosses and whiney co-workers. Interestingly, both good guys and bad guys form their identities in the same manner, discovering ways to feel good about themselves by exploiting the opportunities that surround them. Although we tend to think we control our identities and make our own choices, our environments exert far more influence over who we are than our internal compasses. We choose from the identities available to us. A recent article in *American Psychologist* argued that from a mental health perspective, there is no difference between a freedom fighter and a terrorist except for whose side of the war they are on.[96]

Identities are always negotiated within our larger society, sometimes leading to massive shifts in perception. Think about the legitimacy that lesbians, gays, bisexuals and transgender people now experience in many Western democracies and the celebrations that have followed, like Pride parades and award-winning movies. Though still encountering too much resistance, a young person who comes out as gay these days can enjoy an enormous amount of support in his or her community, and those who do will be much more resilient individuals.[97] We should not forget our changing social environment is what has made it possible for people to identify as part of the LGBTQ rainbow and still feel accepted and powerful.

Because identities are reinforced by appreciative audiences, the best identities are those that are socially approved. Employers cue employees on how to behave. When workers break with convention, they are rarely applauded—unless they happen to save the company or score a win for investors. This is the experience of every adventuresome immigrant who tells her parents that she does not want to stay on the family farm. She will have to ask herself if the risk of being cast as an outsider is worth the potential benefits of forging the identity she wants. It is a fine line to walk between groupthink and individuality. People who assert their individuality are catalysts for growth. On the other hand, risk-averse middle managers get to their comfortable positions because they know not to threaten the culture of their organization or the authority of senior management.

Such is the paradox of identity. Being the same as everyone else makes us resilient. It helps us find friends, maintain employment, and be respected for the little contributions we make. We are part of the herd. But being different is also a path to resilience. If she gets it right, the farm girl can start a new life in the big city and become the center of her own dynamic community. Either way, to be resilient means to be engaged with others. People withstand life's challenges when they have their families, friends, and communities to rely upon. Relationships are just one piece of the puzzle that provides everything we need, but many decades of longitudinal studies on resilience have shown over and over again that we all do better when, among other resources, we enjoy the advantages that come when someone loves us and we have purpose in our environment.[98]

★ ★ ★

I am often asked, "Are religious people more resilient?" That is, does faith in and a special relationship with a higher power predispose us to success? The question comes with an implicit bias: that a higher power exists and that religion confers on believers' spiritual advantages. A dispassionate look at religion and resilience tells a more complicated story.

When people are asked how they have managed to overcome great adversity, like the death of a child, the loss of a home to flooding, or a long war, they often say that it was their faith that carried them through. Yet, if we actually look at what happened in their lives, we see that *faith* is a word that captures a larger set of processes at play. Let me provide an example, and you can decide for yourself if this is true.

A doctoral student of mine worked for a decade with children dying of cancer. I think he was traumatized by the entire experience, but his faith had convinced him that there was a higher meaning to the grief experienced by young children. A devout Christian, he could recall numerous times that he had seen children and their parents praying together. In his research, he talked to children about

death. He searched for the children's understanding of spirituality. He had them draw pictures of their families and themselves.
He asked them about God and meaning, and what they thought
about being sick. In the end, he developed a theory of children's
understanding of the divine.

Although I shepherded him through his graduate degree, I never
felt his study was documenting the real phenomenon that he was
observing. My student had simply interpreted what he heard as
spirituality. What I heard in the children's words were much simpler stories of resilience. They took pleasure from their lives as best
they could. They liked to receive and give affection. They missed the
children in the beds next to them after they died. They could not
really understand death, except as the absence of pain. They missed
being able to play rough. They liked the clowns that came onto the
pediatric oncology unit but hated the rules and the tubes and the
many other invasive procedures to which they were subjected. Did
they have a special connection with the divine, or was it a more
mundane desire to cope as best they could? That was a debate we
never resolved.

During a crisis, many of us will feel that our lives are part of
something much bigger and that there is an unknown purpose to
our existence. We see both our suffering and our success as expressions of our faith. In that sense, faith can help us cope with adversity,
but I doubt God is any better at creating resilience than the politicians who put in place the taxation system that builds hospitals for
sick children, or the medical staff who use the latest technologies
to ease a child's pain. In other words, we give God credit for human
philanthropy and civic action.

Why do people prefer to attribute their own capacity to cope to
God rather than to social policies or the pharmaceutical companies
that make their medications? Why do people relate their personal
success, at least in part, to their levels of devotion? Perhaps for the

same reason that all presidential candidates in the United States have to be seen in church, and to end every speech with "God Bless America." Religion provides us with a common vision: it cements our social networks; it tells us we are blessed and therefore privileged and powerful. In our surrender to a higher power, we paradoxically feel that we have more control over our lives, not less. After all, we have *chosen* to surrender. We have *chosen* to abide by the rules prescribed by our religious leaders. We tithe, or donate, or prostrate; cover our heads, faces, and legs. We eat this but don't eat that. We work on this day but not that one. And we tell ourselves that our interpretation of God is the right one.

Adhering to a religion is a great way to feel unique and powerful, especially during our darker moments. We regain a measure of control and self-respect. To be religious is to resist the randomness of disaster that threatens our existence each and every day. If religion helps people be more resilient, it is because it makes accessible the resources we need to succeed in a crisis.

But does religion actually deliver resources better than, say, a group of friends gathering to watch Monday Night Football or a quilting bee? Yes, probably: religious institutions give far more structure and meaning. They provide a detailed script for how to live well, and they create large networks of support. They provide a sense of continuity over time to our cultural identities. They make us feel like we belong somewhere. When we are treated unfairly, they remind us of our intrinsic worth. For these reasons, religious people appear to be more resilient when bad things happen.

Unfortunately, the connection between religion and resilience is fraught with problems. Some individuals with histories of criminal behavior, drug addictions, family violence, and social isolation turn to religion as an escape from their desperate situations.[99] For these individuals, religion becomes a corrupting influence, where rules are rigid and reason is suspended. It becomes a place where a short passage in a book written hundreds of years ago can be used to justify horrific acts of violence. For these zealots, religion reinforces psychopathology rather than resilience.

The important question is whether or not organized spiritual practices do more good than harm. They are a wonderful crutch for our untamed, rambling, self-defeating minds. They provide wounded souls who are shuffling through our communities a place to reflect on their misfortune. Though I am realistic about what religion can offer people, on the whole, I prefer to see people comfort themselves with self-flagellation and ritualized prayer than destroying themselves with drugs or sinking into a deep depression because life feels pointless. Misused, religion becomes a dangerous weapon that excludes those who are different and justifies unimaginable atrocities. Used well, religion can be a salve over deep emotional wounds—used well, it heals.

Regardless of one's religious beliefs, resilience requires a sense of belonging, purpose, and meaning. Belonging is more than attachments to individuals; it is a sense of connection to a group, an idea, a culture, or a nation. Good workplaces can emphasize engagement just as much as religious congregations. David Zinger, a former executive with the whisky maker Seagram, likes to say that we know we are engaged when we experience "good work done well with others every day."[100] Engagement means satisfaction. An engaged worker finds through his work the resources necessary for well-being and a sense of purpose in life. Create that engagement before a crisis, whether on the job or at home and, as Zinger reminds us, "Work can make you well."

Whether through religion or work or ethnicity, we need this feeling of attachment. An immigrant from Lebanon, Italy, China, Australia, or any of dozens of other countries will find groups of people like himself in every big city around the world. We all gravitate to our tribes. We need our tribes. Where I live, Shambhala

Buddhists have invented Children's Day, held on the winter sol-
stice, to give themselves a place to be while most everyone else is
preparing for Christmas.

Being in close proximity to others also improves our chances
of survival, buoys our flagging spirits, and connects us to necessary information and resources during and after a crisis.[101] George
Bonanno and a large group of co-researchers based at New York
University's School of Medicine undertook a multi-year study of
post-traumatic stress disorder (PTSD) and resilience among US
military personnel deployed to Afghanistan and Iraq starting in
2001.[102] Soldiers were asked to self-report their experience of mental health at the outset and again at three-year intervals. Both the
3,393 soldiers who deployed just once and the 4,394 soldiers who
deployed more than once showed remarkably similar mental health
outcomes. The vast majority, more than 80%, reported few if any
symptoms of PTSD over a six-year period, regardless of the number
of deployments. About 11% showed decreased PTSD symptoms,
while one in twenty described increasing symptoms of trauma
over the same period. One in twenty soldiers with increasing levels
of trauma is not good news, but the important message here is
why so many soldiers with high levels of exposure to *potentially*
traumatizing events coped well during and after deployment.

Bonanno suggests that the outcomes could stem from the sense
of social cohesion within a soldier's unit. They could also be the
result of a sense of mission among the soldiers. And, oddly enough,
they may be related to how much the soldiers smoked and drank
once they come back from overseas. The soldiers who saw the most
action and presented with the least symptoms were those with the
worst health regimes. It is difficult to know why that is the case, but
maybe the soldiers were self-medicating with beer and cigarettes,
or maybe these habits allowed soldiers to maintain strong ties with
colleagues (assuming they were drinking and smoking with other

veterans). If you are going to maintain a sense of cohesion with your unit after deployment, huddling at a bar seems like a reasonable strategy. Whatever the reason, most soldiers do not suffer PTSD in large part because of the protective function of social cohesion.

If you have ever seen a large herd of zebras in the wild, you know that their stripes confuse predators. Picking off and killing just one is difficult when there is such a mass of stripes and hooves. We are like those zebras, all standing next to each other to look more intimidating or less vulnerable than we would as individuals. We feel stronger than we would feel on our own. Think of precision marching or the ritual of collective prayer in a mosque, and you can see the strength in unity. When we all do the same thing, we look, like zebras, much more certain of our tenure in this world and proud of the space we occupy. (The blur of vertical lines is not, however, the reason zebras have stripes, however much they aid survival. Zoologists say the real advantage of the pattern on a zebra's hide is that it causes a cooling effect as air passes over the skin, making it easier to survive a hot climate).

Of course, there is a flipside to group cohesion, which is that tribalism leads to xenophobia, groupthink, and cliques. These conditions do not support resilience for long. When our jobs are being threatened, unions become more important but also self-serving. My workplace is a good example: while I support my union and am thankful for the power of collective bargaining, it is disappointing that it has fought to remove mandatory retirement or performance review for older professors. We have become a tribe that looks after itself rather than one of emerging scholars who need opportunities to start their careers. Full professors with good pensions can tie up spaces at universities and deny jobs to new graduates without any accountability to their directors and deans. We do this under the guise of "rights," but like every cohesive and powerful group, our goal is as much to look after ourselves as to promote a greater good.

This is where resilience becomes complicated. When we use our collective power to fight injustice or survive adversity, belonging is a force for good. The workplace occupational health and safety committee is an example of healthy resistance and well-placed advocacy. Together, people keep each other safe. But we also see dysfunctional, dangerous ways to be resilient in the workplace and in our communities. There are the coffee-room cliques that bully and intimidate, and there are street gangs and informal networks of racists who find security in their homogeneous communities. For them, violence and hatred are calling cards. In my experience, people want tribes, and they will ask to be part of whichever nearby group brings the most power and the best chance of survival when bad things happen.

This is a lesson we can learn from violent extremists. As a resilience researcher, I have been less interested in why people become violent than in why more people do not act out violently, given the manifest injustices in the world. The answer is complicated, but it seems to have a great deal to do with the available resources. An adolescent or young adult who can safely say he has access to all 12 of the resilience-promoting experiences I described in Chapter 1 is far less likely to fall prey to those who exploit young people's vulnerabilities and their search for a place to belong.[103] Despite the ever-increasing global threat of terror, I am an optimist that things will change for the better, and very soon. After all, even the potential suicide bomber will join a legitimate army if given the opportunity. The more we stabilize failed states and give young people a sense of power, the more likely we are to avoid birthing the hatred that fuels suicide-vested children to blow themselves up in service to some cultural anachronism.

This same optimism can be turned on other social issues. Problems can diminish in importance when people are provided with resources tailored to their needs. The abusive alcoholic will participate in a treatment group if that group is easily available, or if an alcoholic's family forces her into treatment. A social misfit who spends his days online might accept an invitation to play Dungeons and Dragons face-to-face with a group of gamers. When given the opportunity for more socially desirable affiliations, we tend to prefer the relationships that offer the greatest power and widest social acceptance. Our resilience depends on us finding and sharing common purpose. When we succeed, we not only have a phalanx of allies to help weather a crisis; we also assert a collective powerful identity. If we feel attacked, knowing who we are will be critical to survival. I need to be able to say, "I am a good worker," "I am the best manager," "I am a devout Sikh (Christian, or Muslim)," or "I am a successful social worker (farmer or soldier)." A sense of belonging and a sense of identity feed on each other, as do relationships, routines, and our sense of control. If resilience is relatively common, it is because there are so many ways we stay connected with others during good times and bad.

Chapter 6

Does Work Work for You?

DOES ANYONE REALLY BENEFIT from those trendy team development retreats most of us loathe to attend at our workplaces? Boil down the sound bites from gurus of workplace health and their advice is simple: "Do more of one thing and less of something else." On the "more" side are instructions to meditate and practice mindfulness, eat healthy, especially at snack time, move more (even if it is just walking), exercise a few times a week, get enough sleep, find a hobby, volunteer, upgrade skills, be assertive, pester yourself with self-affirmations, and work smarter, not harder. On the "less" side are things that contribute to burnout: caffeine, alcohol, recreational drugs, taking your work home, checking your cell phones and computers at all hours. If all else fails, quit your job or find a new career.

Scott Miller, Mark Hubble, and Françoise Mathieu are experts on workplace stress and burnout who share a cynicism toward organizational consultants and those ubiquitous staff development workshops.[104] They find serious flaws with these strategies: most obvious is that they put all the responsibility for burnout on the shoulders of workers rather than employers. The workplace health industry is a way for employers to appear to meet the needs of staff without making any serious changes to working conditions or workplace culture. And it does not work or, at best, seldom works.

Workers who have to change themselves still eventually burn out. Every serious look at workplace stress has found that when we try and influence workers' problems in isolation, little change happens.[105] Most telling, when individual solutions are promoted in workplaces where supervisors do not support their workers (as sometimes occurs in the military and other highly structured workplaces), resilience training may actually make matters worse, not better. Simply put, individual strategies to accommodate workplace stress will not succeed as long as mountains of paperwork, unrealistic deadlines, understaffing, job insecurity, poorly maintained facilities, and administrative incompetence continue. None of these things are under the control of individual workers, and yet they make a greater contribution to worker burnout than a worker's individual beliefs or behaviors.

A better, more effective way to ensure that workplace health activities produce a strong return on investment is to think less about what is inside employees and more about the services and supports workers need to be their best.[106] These can range from access to counseling and coaching to healthier cafeteria food and on-site fitness centers. When employers create positive social and physical ecologies in the workplace, their chances of enhancing their employees' health and productivity improves as well. If it takes a retreat day to identify these needed changes, that is a fine investment of human and financial capital. At least then the responsibility for change is shared between those on the front lines and those in the boardrooms.

Before I am accused of pandering to lazy and unmotivated workers, let me say that there is nothing inherently wrong with a dynamic, fast-paced workplace that pushes employees to produce.

During times of growth, it is reasonable to ask employees to step up and do more with less. But resilience during times like these begins months before a business expands. Administrative systems should already be in place to help people stay focused on their tasks. Employers then will be perceived as competent, or at the least, employees will understand that the rapid growth of the business is good for everyone.

There is a vast difference between experiences of good stress and bad stress in the workplace. Good stress is the pressure that comes with realizing a goal we want to achieve; bad stress is excessive expectations placed on us by a failing environment. Bad stress is the result of needless regulation that thwarts innovation, office politics that cause people to psychologically withdraw or feel bullied, and emotional labor that goes unrecognized, such as the endless demands on retail staff to appease angry customers. Miller and his colleagues remind us that our mental health does not depend on how demanding our jobs are but whether we feel in control of what we are doing and can exercise a degree of personal agency. Add a chance to take a bow and get a little applause from our co-workers in the staff room, and I would guess that most of us would like our jobs a lot more regardless of the size of our paychecks.

This can be particularly important for people who work in industries where the pay is fixed and their motivation to be productive is driven mostly by internal factors. Social workers, health care providers, artists, circus performers, even politicians are often motivated by the meaningfulness of their work rather than the remuneration. To experience themselves as contributing, and therefore empowered, they rely on their audience to reflect their importance back to them. To think that intrinsically motivated people are all positive thinkers who work independently would be a mistake. Like other workers, they need to work in an environment that is tailored to their needs.

A resource that is helpful to a vulnerable worker may have no impact or even potentially be harmful to a worker who feels safe and secure. Take, for example, a talented, creative employee who enjoys the perquisites of her talent. She is respected by others, well-compensated for her contribution, and abundantly capable at her job. Her world can seem enchanted because she is a well-resourced individual. Now imagine we place this low-risk individual into a stifling, routinized workplace, where she is expected to punch in and punch out on time, relationships are hierarchical, and she is provided performance appraisals from supervisors who are not as specialized as the people they supervise. Creative individuals forced to work in uninspiring structures will become frustrated and less productive. Likewise, place an insecure individual into a workplace without boundaries and she, too, will become a problem employee.

Creativity seems especially responsive to context. It has been shown that individuals with sensitive personalities—those who respond quickly to subtle changes in their environment—tend to be more creative when their lives are stable and predictable.[107] They flourish because they are on solid foundations and have the security they need to explore random thoughts and feelings. In the field of resilience, we describe these people as orchids, lovely flowers that only bloom in optimal conditions.[108] Put them in a harsh environment, and they wilt.

Personally, I like dandelions: hardy little flowers that thrive under harsh conditions even if they do not produce stunning shapes or colors. What they do well is continue to multiply even in tough conditions. It is important to know if one is an orchid or a dandelion—or maybe a lupin or chrysanthemum? Different species of flower find different conditions optimal, much like us more complex humans. Regardless of who we are, the trick to success is to match our environment to our individual needs in ways that address the challenges we face. Anyone who says he knows what *everyone* needs is either dishonest or nearsighted.

The workplace is no different from the rest of the world: it is easier and more effective to change environments than it is to change people. Consider the plight of unemployed fishers off Canada's east coast. They are waiting for different levels of governments to ease up on environmental protection laws and fishing quotas so they can work again. The pattern is the same around the world. Fishers may think of themselves as dandelions, but they have proven to be more like orchids in their vulnerability to changing economic circumstances. Irresponsible, opportunistic politicians tell them that they will be able to return to a way of life that has served them for centuries but is no longer sustainable. We know this because over the past 50 years, populations of commercially important species of fish have been reduced by 90%.[109]

Real change in these industries will need to be radical. Christophe Béné, a senior policy expert with the International Center for Tropical Agriculture, argues that we need to drop our prejudices toward people working in difficult industries like fishing and coal mining. They are not incompetent, backward people stuck in last-resort jobs. They should not be blamed for the difficult circumstances in which they live.[110] Béné suggests that if fishers are mostly poor and lack the resilience to make changes to their lives, it is likely because they lack the organization and political voice to get heard, the chance to benefit from a changing economy that has made their sweat and toil obsolete, and the social safety net—education, health care, jobs—to cope with change.

If a community that depends on a resource-based industry wants to succeed in a new economy, it is going to have to find ways to solve these problems. It does not surprise me that political populism is flourishing in places that depend on sunset industries like coal, fishing, and manufacturing. Enhancing political voice is the first thing that needs to happen to help people survive during an economic downturn. How we vote matters more now than it has in decades. Unions certainly help, too, or at least they can be of use when their focus is on the big picture and long-term goals. It strikes me that right-leaning politicians and big unions spend most of their time looking backwards instead of forward. The "days of old" are not a model for a vibrant economy in the years to come. They offer neither leadership nor a strategy to ensure employment a decade from now.

Fishers are poor because their industry is mechanizing and automating, pushing people out faster than we can create good-paying alternatives. Even if fishing stocks return, the days of small fishing vessels catching enough to support a family are long gone. The issue with workplace modernization is not the increased productivity or job loss; it is that people who are being displaced are not benefiting from the changes to their industries. One could say that miners are being exploited much like 18th-century Scottish sharecroppers who were pushed off their land to make way for landowners to graze sheep. With each economic change, the fair and just thing would be to have the wealth generated by the "next big thing" make the lives of those who are displaced easier. Labor-intensive jobs with good salaries are never returning to the United States, Canada, or Europe. They have been usurped by robots and cheap foreign labor.

So do we tax the robots or shorten the workweek to 30 hours? These ideas may seem radical, but so, at one time, was using an income tax to finance a war, or introducing the 40-hour workweek. Fishers might want to demand a dividend on their legacy: they are part of a once important industry, and they deserve some financial support to get ready for new realities; they require resources to transition to new industries and maintain their well-being during this period of disruption; they need assistance with retraining and retooling; they need access to education and incentives to relocate.

If this is taking it too far, they might at the very least see a rebalancing of the income tax burden from those on the bottom rungs of the economic ladder, whose incomes have been stagnant since the 1980s, to those on the top rungs, who have experienced huge growth in their incomes.

★ ★ ★

Paul was certain he was one of the lucky ones. Straight out of high school, he found a job on the overnight production line at the General Motors plant an hour from where he lived.[111] It was excellent money at a unionized shop. He thought he had a job for life with the world's largest automaker. Weekends were easy. He was flush with cash even after he married Tina and they bought a small bungalow in the new subdivision out by the lake. Paul bought a new car every few years. Tina drove the old one, which was fine by her because their three kids were always messing it up on the way to sports practice. Paul earned promotions that provided him with more training and a fatter paycheck. The family weathered the high-interest rates of the 1980s and ignored the dot-com fiasco of the 1990s.

Early in 2006, GM announced that its cars were not selling, and 30,000 jobs would be cut. Entire factories were closed. Overnight, Paul, in mid-career, was out of work. He had a year to retool. There was not much prospect of finding a job close to where he lived with his qualifications—and certainly not at his previous level of pay. There were jobs out West, however. That same price jump in oil that had hurt GM's big cars had fueled the energy sector. It looked like Paul and the family would have to move, or else Paul was going to have to move and see Tina and the kids two weeks out of every five.

His oldest daughter, Charmaine, watched her father struggle with the decision. She was supposed to head to college the following year. She had been thinking about how important it was to make the right career choice. But with her father's layoff, she was beginning to wonder whether getting it right was even possible. "What's the point of planning at all?" Charmaine, like Paul, was rethinking the promise of certainty that any career path brings.

Employers, counselors, and even politicians are still trying to convince us of career myths. Recent elections have promoted head-in-the-sand economics that ignores the global movement of capital and labor. We want to protect our industries at home, but who is going to buy a $1,000 television when the same item built in Vietnam can cost $400? We tell people there will be good jobs at fair wages, promising advancement and security for life. People believe these myths. They convince themselves a stable career path will follow if they do everything they should. Good marks, a good education, good training, an entry-level position, impressing the boss, improving credentials, and voilà: a career, a home, a family, a car, a dog, and two kids. Unfortunately, I am meeting fewer and fewer people on this path to success.

We want to believe that there is one best career for each of us and we need only to discover what it is. Yet when Deborah Betsworth and Jo-Ida Hansen interviewed 237 older adults in the US, they found that chance events were a ubiquitous part of their lives.[112] To think careers are built with intention and design is naive. They are shaped, instead, by the many resources life puts in front of us—chance encounters with what become personal areas of interest. These include personal connections that tell us which jobs have a higher status, might interest us, or are easily acquired, and professional connections that help us gain an entry-level position or gain an advantage over other applicants, as well as unexpected advancements because someone above us has been fired, has been injured, or has quit. Sometimes it comes down to being in the right place at the right time, or to random encouragements from people we did not even know were watching us. Our need to balance family commitments and our spouse's career choices also mess with our plans, as do sudden changes in the job market, making last week's hot career choice obsolete. Likewise, changes in political agendas mean more or less money for research and development, or retraining allowances, or shaking up the public sector job market.

What is fascinating about this list of disruptions is that it reminds us that our success relies on circumstances over which we have little or no control. It no longer makes sense to think that what one chooses to do early in life will be how one earns an income a decade later. Careers today are more like our clothing choices than tattoos, their durability measured in months instead of years. Experts in training and personal development tell us that those who thrive in the new economy are those who are energized by change and can tolerate and manage ambiguity, and who have relatively little need for control.[113] They also have talents that are valued by employers and the generic skills required to get and hold a job. Self-reflection can help but does little if the economy is in recession. No amount of personal growth or individual skill is going to improve employment prospects when there are no jobs to be found. In a floundering economy, the only way to be more resilient is by retraining, relocating, or altogether reconsidering our life design.

Fortunately, there are many stories of career resilience to balance the despair one feels when meeting Paul. For example, my friend Luis, now in his mid-50s, did as he was expected and attended university. He studied biology, thought about medical school, finished his Bachelor of Science degree, and instead was certified as a surveyor. When that became boring, he went to community college and became an auto mechanic. He loved it, at least for a few years. Eventually, he transferred his skills from the routine of the shop floor to the manager's office. This was a clever strategy to find a greater challenge and still be connected to fixing cars, which he loved. A growing family and a mortgage affirmed his decision to seek an office job. Luis was ahead of his time when it came to recognizing that career paths are fluid, yet even he could not avoid the obsolescence he experienced at age 54, when the service department at which he worked laid him off permanently. Still, Luis took the situation in stride. He spent a little time at home, then discovered that he could sell cars as easily as he could fix them. His job as a used car salesman does not pay as much as he earned as a manager, but his bills are not as high, either, with his kids now out on their own. All in all, Luis has had a good run, though his career path has been anything but a straight line.

I can relate to Luis' plight. In the 34 years since I graduated with my first degree, I have been a journalist, a community worker, a child and youth care worker, a social worker, a family therapist, a woodworker, a professor, a researcher, a writer, and a small business owner, often managing several of these roles at the same time. Like Luis, this amble through careers has given me many skills that have proven useful in many different job markets.

These days, career flexibility is not optional. Even teachers are no longer restricted to the core subjects for which they were trained; nowadays, they are also mental health workers in the classroom. They teach math, prevent bullying, watch for signs of suicide, build children's self-esteem, and talk about sex. Instead of giving grades, many write motivational notes to students and parents describing the child's progress and "unrealized potential" rather than assigning a D+. No surprise, then, that the most successful among us have a depth of knowledge in one area and a lot of other skills besides.[114] That is what makes the stay-at-home mother of three a far better pick as supervisor at the local fast food joint than the 25-year-old with an MBA. Resilience is all about flexibility and having opportunities to realize our potential. Business knows this. We are witnessing unprecedented changes to how companies adapt to economic shifts. Waste disposal companies have morphed into green energy innovators. Bankers have turned into money managers. Pulp and paper workers are growing sustainable forests. Even our military personnel are being asked to be peacekeepers as often as they are warriors. The mantra of our time is flexibility, the same flexibility that allows us to take advantage of new opportunities as they arise.

The career myth may be dead, but the science of career resilience is only getting started. So far, what we know is that our careers, like our relationships with our friends, colleagues, spouses, and even our parents, change over time. At each juncture, we will need a set of coping strategies if we are to weather change and come out stronger. Manuel London and Raymond Noe, professors of management and human resources, defined *career resilience* as "the ability to adapt to changing circumstances, even when the circumstances are discouraging or disruptive."[115] When change is afoot, we need to bend, not break. We need to let go of the outdated and embrace the new. We need to accept that life will be full of disappointment and uncertainty, and that our resilience depends on our ability to adapt.[116]

Like other parts of our lives, personal disposition is important to career resilience—but only to a point. While we need to sharpen the saw, as the Buddhists say, and ensure we are at our peak performance, we also need a large stack of firewood out behind the house to keep us warm while we look for work. Studies of career resilience that are sensitive to the economic conditions in which people live have found that resilience is more likely when social and physical resources are plentiful.[117] Personal fortitude seldom overcomes a catastrophic economic crisis. After all, unregulated bankers were the ones during the Great Recession of 2008 who were responsible for falling house prices and job losses, not the individuals who suffered the consequences of the mortgage lender's stupidity and avarice.

A proper focus on career resilience redirects attention from individual career failure to anticipate disasters before they happen and addressing the social and economic factors that cause individual vulnerability in the first place. We need only listen to the likes of Jim Keravala, co-founder and CEO of OffWorld, a robotics development and mining company, to know that the future of work and leisure is going to change remarkably.[118] OffWorld is gearing up to put robots on Mars and asteroids to make way for human settlement and mining operations. Keravala talks of synthetic biology that will make genetic engineering so commonplace that we lose sight of what is gene manipulation and what is naturally produced. He lauds the potential of graphene, a semimetal formed with carbon atoms arranged in a hexagonal pattern, which has the ability to create materials with 100 times the strength of steel, and with better insulating and conducting capacities, as well. Artificial intelligence and more efficient forms of energy, he says, will bring dramatic changes to how and where we manufacture goods. The combined result of these changes will be a tectonic shift in the labor market in just one or two generations, with a huge need for more sophisticated skilled labor with postsecondary training in the STEM (science, technology, engineering, and mathematics) subjects. Or maybe not: artificial intelligence may take care of the need for talented engineers.

Keravala is so certain of the coming disruption that his company has started a fund to share the anticipated billions of dollars in profits it will earn with populations affected by new technologies. He is not alone in predicting the displacement millions of people who currently work as truck drivers, personal care workers, and miners. Tony Seba, author of *Clean Disruption in Energy and Transportation*, believes we are just a few years, not decades, from great disruptions in how we move ourselves and our things around, and create the energy we need.[119] To Seba, the coming changes are as predictable as the replacement of horse-drawn carriages with automobiles and Kodak film with digital camera technology. He asks us to consider why smart people at smart organizations consistently fail to predict market disruptions. It is a good question. Resilience in the workplace may be a survival skill we need much sooner than we thought, unless we are content with the prospect of millions of unemployed drinking themselves into oblivion from boredom and depression, or electing authoritarian leaders who con them into believing that the past was better than the present.

If all of the future-workplace predictions come true, the resilient workforce of the future may actually be that which is gainfully *unemployed*. Keravala thinks the economy of the future will be measured by how few people work rather than by how many. A 6% employment rate may signify a developed economy with the capacity to automate, while a 26% employment rate might be seen as backward. The question then becomes what do we do with all that surplus human capital? The answer is far from clear, but if resilience is more what we receive than what we have, society will have to reorganize to provide every adult with two critically important opportunities. First, we will need to have a plan for the redistribution of wealth so that people are economically secure even if they are working few hours each week. The second, far more dangerous challenge will be finding ways to provide opportunities for people to make a contribution somehow, someplace. If we do not solve these problems, we may be faced with hordes of displaced, angry young people who have nothing meaningful to do with their leisure time and who are shut out from the economic benefits of an automated society. If history teaches us anything, it is that unemployed youth (especially young men) can quickly destabilize societies.

Given that there is not much that individual workers can do when their call center is moved from Memphis to Mumbai, or their steel factory goes robotic or to Mars, a good career development strategy is one that acknowledges constraints and promotes resilience by facilitating access to required resources; among the most important is education.[120] We cannot be fooled into believing our undergraduate credits are going to bring us a permanent job. Lifelong learning is no longer optional. Over 80% of people who complete an undergraduate degree will enrol in further post-secondary education, often at vocational colleges where practical skills are taught. If changing workplaces are inevitable, then it will be individuals with the most diverse training and experience who will be able to choose their career paths. Flexible credentials and a willingness to try new things improve our chances of success.

What about passion? How important is it to love what you do? Career specialist Barbara Sher has said that all great human endeavors begin with a wish and a dream.[121] Wishing alone will not make things happen, but without some future orientation we are liable to end up unhappy and unproductive. Add external resources like money and training to make the future happen, and we can envision possibilities, no matter how far off on the horizon. Passion, however, is not everything.

Years ago, I met a young man in my clinical practice at a community mental health clinic who was starting a mechanical engineering program at university. Anastasios was doing fine academically, but he had little interest in the subject matter, which was heavily weighted toward math and physics. He had never liked those subjects, even in high school. He passion was woodwork. He was great with his hands, but his parents had dismissed his talent as an unrealistic foundation for a career. They refused to pay for him to go to a vocational college when he had the academic talent to attend a university, and they had the financial means to pay for it. A hardworking pair, Anastasios's parents had clawed their way into the middle class. They were not about to watch their son forego a brighter future to make funny-looking wooden tables. Anastasios compromised with his parents and went into mechanical engineering. He promised himself he would get his degree and then return to college to become a woodworker.

Anastasios's story is remarkably common. He had to compromise, but it was probably the right decision in the long run. Resilient individuals tolerate the discomfort that accompanies their search for work and seek a balance between doing what they are good at and what will support them. Sean Aiken performed a different job every week for a year to better understand what makes our time at work fulfilling.[122] Aiken found that passion on the job was not the best basis for long-term happiness. It is the people we work with and a feeling that our work is significant, no matter what we do, that makes our time at work satisfying. Once again, environments count far more than individual motivation. Given these findings, I am inclined to advise individuals like Anastasios to hedge their bets. Get the training needed for financial sustainability, then put aside time to follow passions. In the long run, finding a perfect life is about creating a balance between what we can earn and what we want to accomplish.

I recently met in my office a professional photographer named Jessica, who thought she could make it on her own when her job at the *Daily News* ended. She welcomed the opportunity for change. She had always dreamed of becoming a landscape photographer and publishing a book; newspaper work was what paid the bills. Years of running to crime scenes had honed her ability to shoot, day and night, in any weather, but she had never found much satisfaction in it. Once her paycheck ended, however, it soon became clear that her passion for landscapes was not immediately going to put food on her table. Building a reputation and portfolio would take years. Given her options, Jessica did what most photographers do: she shot weddings. She did well at it and made good money, and the landscape idea once again took a back seat. It was not long before she was surly with customers and on the verge of blowing her reputation as a photographer and losing everything she had built for herself.

The solution was balance. Jessica came to counseling angry with the way she was messing up. She left with a simple plan: a commitment to devote one week each month to landscape work. The weddings would just have to wait. Within a year, she had a portfolio of material, and she had made her first sales to two different magazines. "Next year," she said, "I'm going to shoot landscapes half-time. I won't earn as much, but I really think the extra week a month will make a difference."

Doing what one loves to do for financial security can kill the very passion that motivated us in the first place. Individuals like Anastasios and Jessica who demonstrate career resilience seem to understand that how we earn money and how we express what is special about us can occur in different domains of our lives. A good career and life design work best when they make space for self-expression, on or off the job. In my community, I can name a concert violinist who is a full-time doctor. I know an architect who is a carpenter on green projects. There is a world-class rhythmic gymnastics judge who works as a school administrator, and the best peewee baseball coach in town drives a garbage truck. These are people who live their passions during their free time. I also know people who live their passions through their main occupations: a cellist who loves his full-time gig with the local symphony, a professional trainer of Olympic athletes, a teacher who loves her students, and a municipal worker who is more than willing to get up in the middle of the night every time the sewage system breaks. It does not matter which way you do it. The result is much the same: contentment and an incredible ability to withstand the ups and downs of life; our careers are rewards for finding a way to do something one loves at some point in the day.

Individual choice should not be oversold. Things worked out for Anastasios and Jessica because balancing their needs for a stable income and their passions was realistic and achievable. They may look like rugged individuals, but they had plenty of resources, including education for him and on-the-job photographic training for her. They were able to find opportunities that suited their objectives. Passions only become a curse if one's employment requires a 10-hour workday six days a week, if it pays minimum wage, or if it offers just two weeks of holidays a year. In such situations, there is no time or space to realize fully one's potential.

Nor should passions on the job be oversold, nor the democratic workplace or the indulgent office with free food and foosball. It is easy to find employee retention studies that advise transparency, frequent communications, and accountability.[123] These democratic approaches generally glide over the pitfalls: that already-stressed employees often do not want to share responsibility for every decision the employer makes; that others will view corporate town halls as a waste of time or be disappointed when their suggestions are ignored. Millennial workers are often lured to offices with vegan cafeterias and games rooms, but evidence shows that this supposedly entitled generation wants what everyone else wants: steady employment, recognition, promotions. In one survey, 200 recent grads and undergrads at a business school ranked the factors they considered most important in a job. Their answers were compared to those of a bunch of managers in the 1960s and found to be remarkably similar. Both groups wanted respect and good pay.[124] Forget the foosball; shelve the search for meaning. Decent salaries and opportunities to make a contribution are what count.

★ ★ ★

It is much easier to find a successful career path if that is what we are looking for, when we are open to the serendipity of life. Though research on this topic is scant and inconclusive, I have met individuals all around the world who have shown me that careers flourish when we are open to being disrupted. Once again, it is the well-resourced individual whose life is full of random events who is more likely to live a fulfilling life. A few years ago, I met an enchanting young British woman who produced documentary films for a national broadcaster. Fatima had planned to get a PhD in economics and had thoughts of becoming a university professor. She was studying abroad, completing her master's degree at Columbia University in New York City when everything changed. While on her way to school one day, and feeling particularly homesick, she heard a man in his 50s say something to a newspaper vendor at the entrance to the subway. Catching his accent, she approached him and said, "I am so happy to hear a Brit."

"What he must have thought!" Fatima said when I asked her about the encounter. "I was all of 24 years old. He probably figured I was scamming him. At least he looked wary at first." That changed when she stopped asking him questions about what he was doing in New York and began telling him about herself. He stood there, transfixed, while she spoke. Born of an Indian father and Irish mother, Fatima has engaging eyes and an indeterminate complexion. She is a poster child for multiculturalism. The man she was talking to turned out to be the bureau chief for the BBC. He needed a reporter. He suggested they meet at his office the next day, and the rest is history. Fatima has produced award-winning documentaries for the past decade. She never did finish her master's.

There is more than just luck in Fatima's story. Remember Seneca: luck is what happens when preparation meets opportunity. Take 10 chances, and odds are that something good will happen at least once (especially if you have the resources to survive the other nine). In part, this success can be explained by the momentum for change that our efforts create within our environments. That momentum leads to opportunities to realize our potential. Norm Amundson, a career development expert at the University of British Columbia, uses the principles of physics to help explain the way successful people navigate their career paths.[125] We know, for example, that it takes far more energy to get a car started than it does to change a car's direction. It is the same with people. It is always better to be moving forward with a life goal than simply waiting for Plan A to make us successful. In fact, we are most vulnerable when we are the least flexible and most stationary. Taking chances changes the odds that we will succeed. The world around us tends to appreciate the *chutzpah* we have shown.

Chapter 7

The Institutions We Need (But Seldom Get)

FIFTEEN YEARS AGO, when I began researching resilience, I was invited to Medellín, Colombia, to see how the most violent city on Earth was transforming itself. Home to the infamous drug lord Pablo Escobar (before he was killed), Medellín was riddled with violence as gangs fought for control of the lucrative drug trade. Paramilitaries were at war with the military, and somewhere in the background was a terrorist organization known as FARC. It was a tremendous shame. Medellín is a stunningly beautiful city, set amid the rolling Andes Mountains on the equator. Its climate is constant, with warm days and pleasantly cool evenings all year round. The hills are dotted with greenhouses growing flowers exported to the world. A lucrative lingerie industry adds to the mystique and eroticism of the Antioquian people.

During my first trip to Colombia, the streets were so violent that I remained hidden in the offices of the University of Antioquia, with armed guards at the gate. I could not venture into the favelas that ring the city, climbing up the mountainsides in progressive waves. But I did meet many courageous educators and youth workers who had devoted their lives to break the cycle of violence and to help children find futures that did not include drugs, guns, or gangs. Among the most inspiring was the principal at one of the P–12 schools near to where Escobar used to live. She stood less than five feet tall but commanded everyone's respect. She kept her school open, even after stray bullets from street fights had killed four students on her playground. "Where else would the children go?" she asked me, not expecting an answer, but also not expecting my admiration. "In my community, when children hear gunfire in the streets, their parents tell them to hide behind cement walls, then carry on to school when the gunfire stops." She must have seen the shock on my face. "They have adjusted," she explained. "They had to."

Fortunately, the violence in Medellín has subsided. Better policing helped, but so too did changes to the infrastructure of the city. People were given hope through substantial public investment in their futures. Mass transit and access to jobs, libraries, and improved public education signaled to people that there were opportunities for them and their children beyond the drug trade. Programs for children, parents, and educators challenged everyone to break the cycle of violence. Teachers were taught to discipline children without resorting to corporeal punishment, and parents were given the support of in-home counselors to help them do the same. Children that had simply accepted violence as a normal part of their culture were learning to expect fair treatment from caregivers.

Resilience can be mere theory until you witness an entire population becoming more successful. I made many visits to Medellín and each time noticed that my movements were becoming freer. I could walk to a mall or spend time in the downtown square during daylight hours. Eventually, I visited the school where that principal worked and saw her walk among her students. They followed her like a comet, a tail of bright faces waiting to be noticed. And she did notice them. She leaned down (or reached up) and called each child by name, kissed each on the cheek, and gave them a gentle pat on the bottom to send them on their way. It was clear to me then why those children chose school over life on the street. At school, they found hope and relationships. At school, their future was tangible, real enough to make dodging bullets worthwhile.

Whether I look at the resilience of the children or the resilience of their educators, I come to the same conclusion. Complex problems need complex solutions that affect multiple systems. Our collective well-being has improved over the past century largely as a consequence of social investment, good governance, and advances in public health. None of this depends on partisan politics. When it comes to social change and resilience, I am an ideological pragmatist. I care little which party is in power, as long as the politicians have the humility to listen to experts and are willing to do what is necessary for long-term success. If that means building mass transit systems, then build them. If that means including a profit motive in health care, then include it. If that means providing employee assistance plans for stressed workers, then provide them.

One need not travel to Medellín to find stunning examples of social transformation; they can be found in almost every part of the globe. The area around the old locks in Minneapolis has been transformed from a post-apocalyptic industrial wasteland into a vibrant arts, entertainment, and food district. Other cities have done the same. Bilbao built its Guggenheim Museum and in the process revitalized the entire Catalina region after its steel industry failed. London's South Bank has become renowned for its vibrant street life and the London Eye, erasing the industrial blight that existed there a few years earlier. The list of such transformations is long.

When it comes to success in really bad places, nurture and nature are not equal parts of the equation. Nurture trumps nature every time. The implications of this simple truth are profound. With the right incentives, says Richard Thaler, a Nobel laureate in economics, we can "nudge" people to do things that are in their best interest.[126]

Thaler promotes a style of government that limits the power of the state to force its citizens to do anything but makes it easier for us to act in our best interest. Take cigarette smoking. Banning cigarettes outright would likely do more harm than good, but nudging people to stop through higher taxes and limits on smoking in public places is in everyone's long-term interests. This approach works because it leaves people feeling in control of their lives. They "decide" to quit, even if the decision is driven by extrinsic factors. Of course, the feeling of choice is largely illusory. Cigarette smoking is declining in most Western countries, not because individuals have voluntarily dropped the habit but because they have been pushed to live healthier lives by government policy. This is a far more effective approach than trying to convince individuals, one at a time, to make a difficult change.

No two ways about it, resilience is political. It does not require big government, but it needs government or a substitute authority that will provide the foundations for success. Who else will make the investments in education to fuel our postindustrial economy? How else will we ensure that we have sufficient numbers of educated individuals capable of handling jobs in a new economy? Social responsibility and economic success go hand in glove.

Our misplaced faith in individualism is partly to blame for the disrepute in which necessary government actions are often held. To most middle-class, well-educated, financially secure people—lucky folks with lots of advantages and few problems—the institutional supports that make us resilient are invisible. They do not see the privilege that makes their education and employment accessible. They do not see that government institutions like policing and bank regulation have favored them. (For decades, banks in the US redlined communities, making it impossible for people of color to secure credit.) No wonder proponents of individual grit see self-transformation as an endeavor available to anyone who shows up at a self-help seminar. They are speaking from personal experience.

I was reminded of this blind belief in individualism a few years ago while listening to Jon Kabat-Zinn in a Washington ballroom in front of 3,000 psychotherapists. He had us sit quietly with our hands on our knees and experience the rhythm of our breathing, and he talked about mindfulness for 90 minutes. It was restful and a good show; but it was what he did for the last three minutes of his talk that caught my attention. He encouraged us to make the world a better place so that more people could be mindful. He speculated that mindful leaders would seek to end conflict through peaceful solutions instead of war. For just a moment, he seemed to acknowledge the futility of what he had been preaching for the previous 87 minutes. Nobody, however, paid much attention to what he said in his closing remarks. Nor did Kabat-Zinn dwell on what individuals could do to change the world. He suggested that we could get better politicians by encouraging them to breath deeper and meditate.

I understand the problem. Who wants to listen to a motivational speaker tell you that most of your success has very little to do with what you think, feel, or do? That has zero marketability, even if it is the truth. Maybe Kabat-Zinn knows full well that changing people one by one is ridiculously difficult. Maybe he is just taking the path of least resistance. Whatever his motives, I find more to admire in Martin Seligman, whose work is the wellspring for much of what positive psychologists believe. He retracted his support for his earlier work in a 2008 book.[127] He showed real integrity by acknowledging that a tight focus on changing people's thinking disregards the real conditions of people's lives and their lack of access to opportunities and health. Change the institutions that people depend upon, and far more people will benefit.

* * *

All over the world, I meet people who do better with healthy doses of rules, routines, and expectations—in a word, *government*. We all need it, in one form or another. American anti-government zealots are usually keen to substitute some sort of religious control for democratic control. Most revolutionaries have some vague notion of a new regime with which they will replace the old. Even if we prefer to think of ourselves as completely autonomous beings, as lead actors in our own movies, we all do better with order. We are safer and more prosperous, and we live longer, under systems of good government that place some restrictions on individual freedoms, whether they be speed limits or a child's access to cigarettes. We perform better when we stick to schedules, when we are conscious of other's expectations of us, and when our actions are consistently subject to natural consequences.[128]

Rules, routines, and expectations make our lives predictable and secure. This is why children need bedtimes and chartered accountants need job security if both are to weather personal challenges. This is why students do better with clearly defined deadlines and healthy expectations from parents that they pass their exams. This is why we live longer when someone needs us to get out of bed in the morning. Without these structures, people go to incredible lengths to make their lives predictable. They will check themselves into detox; they will marry the first suitable mate that comes along, or they will do something so incredibly stupid that lands them in jail.

Our need for structure increases in times of crisis. A colleague of mine went to a local high school the day after a student committed suicide. What she found were students in the halls wailing, overwhelmed by grief. Everyone, it seemed, had been the boy's best friend. Grief was like a virus infecting the whole student population. That happens when people (of any age) are overwhelmed by new emotions and left to cope on their own. On another occasion, my colleague visited a different school after a boy died of injuries from an accident on an all-terrain vehicle. When she entered the school, the halls were clear. The counselor was directed to the library, which had been reserved as a quiet space for children to come and talk with an adult if the death of their classmate was troubling them. Grief was orderly, and the paths to healing were predictable and well-resourced. In a climate such as this, everyone's mental health was assured.

The last example is the way most communities grieve. In the Jewish faith, people sit Shiva for seven days while family and friends drop in to offer condolences. After seven days, it is expected that the one grieving returns to her normal life, heavy in spirit but with the expectation that it is time to begin to move on. Wakes and funerals do the same thing. They provide structures for us to manage grief at a time when we are understandably overwhelmed.

Structure can also give rise to great creativity. It is as if we must learn to color inside the lines first before we can break the rules. Not all of us, but a surprising number of us benefit from a period of conformity before testing the limits of convention. While I was traveling through France last summer, I went to the Musée d'Orsay and heard stories about impressionists like Monet, Matisse, and Picasso that lend credibility to this theory of conformity before creativity. All three artists were at one point or another kicked out of the Parisian art establishments of their day. Their paintings were judged inferior by those who thought they knew best. What is interesting, however, is that each of these artists began as an aspiring inside-the-lines master before shocking the world with new ways of seeing. These days, dot-com entrepreneurs are some of the most unconventional among us. They are often university dropouts. They do fine without structure, although all of them are hyper-intelligent and capable individuals with firm grasps of the principles of their disciplines. They master the existing digital world before breaking out of their own. Individualism is almost always the prerogative of those who have mastered conformity.

A former client of mine, Alayna, showed me the power and simplicity of structure during chaos. Juggling a part-time job and a young child while completing her master's degree, it was a wonder she had not burned out earlier. When we met, she was frazzled, her hands twitching from the river of caffeine she ingested every day to keep going. Amid tears, she told me she was ready to give up. There were too many demands on her time. It was obvious something had to change.

While Alayna was convinced she should do less, I suggested she continue to do the same amount (she was, after all, highly motivated) but put in place routines to make her days easier and better planned. I could see that she was losing a lot of time each day figuring out what to do next. Before she threw away her job, her degree, or her child, we discussed building into her day a 15-minute planning period, something that she recalled doing in her early twenties when first at university. Back then, Alayna would amble into her favorite coffee shop on the way to school and think about her day. She always enjoyed those few minutes of reflection. It proved to be an easy fix to have Alayna avoid the drive-through after dropping off her daughter at daycare on her way to work. Instead, she would park her car and walk inside for her first coffee. She needed just a few minutes to calm and consider everything she had to get done.

In Alayna's case, the structure appears to be self-imposed, but that would be a shallow read of a much more complex situation. Alayna's need for structure, and the decision to impose more structure on her day worked because her school, her workplace, and her child already placed expectations on Alayna. All she was doing was juggling the pieces. I worry much less about people like Alayna with external demands on them than about people with very little or no structure at all. A week or two of vacation is one thing.

A post-retirement hangover that lasts months because no one expects the retiree to get off the couch is quite different. Resilient people live lives full of expectations from others, even though many of those expectations are self-imposed.

To further examine the value of structure consider a person who has just lost her job. Her credit card is maxed. Life looks pretty awful. In many instances, this person might complicate her mess by making a bunch of changes at once. She blames her spouse for a lack of empathy because she did not get the support she needed to hold the job. For whatever reason, she seeks a divorce or has an affair. She buys something with her last paycheck to make her feel better, ignoring her bills. She parties at night. She stops getting out of bed in the morning. In every imaginable way, she makes herself less resilient by walking away from structure. Soon, she is a victim of the very rules she is trying to avoid. Her marriage is over. She loses her home. Her phone does not work, because the bill has not been paid. Her license has been suspended for driving under the influence. The municipality fines her for not removing snow from her sidewalk. In ways large and small, her life has fallen apart for lack of routine and structure. Her life is punishing her for her nonconformity. This is not the way resilient people handle a crisis.

Now imagine a different scenario. A woman loses her job and her credit card is maxed. Best-case scenario, she has been let go by an employer who provides career planning and financial advice, four sessions of each. She approaches her bank to restructure her household debt. Her relationship with her spouse is bordering on emotional abuse, but she hangs on to it, at least temporarily, for the stability it creates. Her spouse even takes a second job while she looks for a new one. The in-laws pitch in. She sells your late model car and takes a hand-me-down from a relative. All of these things keep her busy, allowing her to make payments and meet expectations. It is reasonable to expect that she will find work within six months, even if she has to commute or relocate for a while. A year later, she has paid off the credit card that she wisely destroyed after she lost her job. She divorces her husband. By waiting, she has made it easier to navigate this difficult transition with ease. The smartest thing we can do in a crisis is fulfil expectations and carry on with as many of our normal routines as possible.

Structure wins over anarchy every time. As citizens, we need laws that reflect a social contract. We agree to adhere to rules in return for safety and freedom, and when we feel our lives sliding into chaos, structure is our salvation. We can seldom influence the laws that govern our behavior, but we can work with them to bring order to our lives. Problem gamblers can ask their local casino to put them on a no-play list. The shopaholic can cut up his credit cards. Forget personal transformation: exploit the structures already in place to serve us. Make the world accommodate our needs, and individual change will follow.

Of course, all those warnings about too much of a good thing apply to structure, as well. Overkill will suffocate resilience and become just as burdensome as anarchy. Overprotective parenting, for example, can be highly dysfunctional. The parent who protects her child from every psychological and physical bump and bruise actually undermines her child's resilience. Anxiety disorders are among the leading reasons for children's emergency room visits and hospitalizations,[129] and they are crippling university students by the tens of thousands. We are witnessing the failure of an entire generation because of the obsessive control of parents who think they can ensure a child's future by controlling every moment of childhood. In some neighborhoods, it is said that the fetus is not considered viable until it graduates medical school. These children are doomed to fail. The first speed bump, the first teacher or boss who says, "you need to perform better," will produce an emotional crisis of epic proportions.

Children need to be free to scrape their knees and make mistakes within predictable environments. This will allow them to come to understand the natural consequences of their actions. In this way, resilience thrives in predictable environments.[130] It is the same for adults. In a parliamentary system, the political leader whose party badly loses an election is expected to resign. In business, a corporate leader whose share price goes into the tank and stays there is soon unemployed. Leaders and the people who work with them do better when the consequences of actions make sense to them. We do better as families, communities, and colleagues when consequences are understood to be in our long-term best interests.[131] When this structure breaks down, and actions and consequences do not match, we become upset. Think about the indignation people felt a decade ago on hearing that Detroit's failing auto executives flew to Washington on private jets to arrange billions in government handouts.

The consequences of actions should be administered in an intuitively obvious and reasonable manner. They should make us uncomfortable while still being proportionate to the problem we caused. When a child will not eat her dinner, and a parent reheats the meal and serves it to her as her bedtime snack, he is giving her a measured, timely, appropriate consequence. There is nothing punitive about this parenting strategy: the food is reheated and the child's need for a healthy meal is met.

The same reasoning can be applied to more serious problems. A man cheats on his wife. Marriage counselors used to encourage offending partners to tell their spouses about their affairs. The problem with that approach was that it did not work.[132] Many marriages held together, but with a stain that could not be washed away. More recently, therapists have been avoiding disclosure and favoring new behaviors. The person who had the affair is encouraged to admit the offense to the therapist in private and then take concrete steps to end the relationship or get a divorce. If the client wants his marriage to continue, a letter must be written and sent to the person he had the affair with, stating clearly that the affair is over. All contact must stop. All that energy that went into the surreptitious activity needs to be put into the marriage. In other words, the consequences are tangible and the solutions achievable. Actions fix the problem, not words. Of course, if the offending spouse does not want to end the affair, or invest in the marriage, then it becomes blindingly obvious that the marriage has ended. At that point, disclosure of the affair makes sense—not to seek forgiveness and redemption but to discuss where both spouses go to from here. All of this comes down to inviting people to fix their mistakes and experience natural consequences.

We actually do better when we surround ourselves with environments that hold us accountable. Good consequences, like healthy amounts of structure and being made to eat our dinners, make our lives more predictable. They also make us far more capable of weathering future crises.

* * *

Resilience is like a game of mirrors. Imagine standing opposite someone, hands raised, your palms nearly touching. You lead first. As you move your right hand, your partner follows with her left. If you smile, she smiles. If you take a step backwards, she does the same. Now swap. Your partner leads. So far, so good. We have established that whoever has the power in a relationship gets to decide what the other person experiences. If I am an employer, and I think like this, I establish rules at work, hire workers, expect them to do as instructed, and pay them for their labor. This is a simple model that works well if one is operating a fish processing plant and the goal is consistency in the size of the fillets that are produced. It may not work so well when the work is dangerous or requires high levels of autonomy and problem-solving skills. For these situations, we need to adapt the mirrors game.

Go back to the first position, your hands raised palm-to-palm with your partner. Only this time, before you start moving, do *not* decide who is going to lead and who is going to follow. Just start and see what happens. If you and your partner are like most people, eventually someone moves, the other person follows, then the follower initiates a new movement and the person who started takes the place of the follower. Sometimes one person starts and just keeps leading. (I joke that these people are bullies, but I really mean zealous leaders). Sometimes neither person moves at all and you both stand there, frozen in awkward silence.

Through both my research and clinical work with people of all ages, I have come to understand resilience as an experience much like this complicated mirrors game. We have the responsibility to tell others: "This is what I need to cope well with the challenges I face in life." We also need to make the best use possible of the opportunities that are given to us. In other words, we need to be able to lead and follow at the same time, directing our lives and letting our lives be directed.

To succeed under stress, we have to be able to navigate to the experiences we need.[133] I like to think of this as the difference between being adrift on an ocean with or without a compass. Personal motivation to set our sails and direct our destiny will be of little use unless we know where to find a safe place to land with food and shelter waiting. All of our personal power will not save us without a facilitative environment and the tools and resources we require. Navigation, however, is only half the story of resilience.

If being a well-resourced individual were enough, I would expect that once a person who struggles with an addiction to alcohol is provided with a treatment program, he would be set for life. But there are far too many exceptions to this sensible proposition. Many of us do not take advantage of the resources within our reach. Anyone who has tried to treat an alcoholic knows that making opportunities accessible is seldom enough. The supports we value and make use of are always a matter of negotiation. As in the mirrors game, alcoholics go to treatment only when they feel compelled to move in that direction—when life becomes so painful that change is thrust upon them.

The same principles of negotiation are at play when some promising young employees frustrate their employers and refuse to accept promotions. Many a young person in both my research and my clinical practice has said that he prefers a lower paying job with less responsibility and more direct contact with clients than a position with more responsibility, higher status, and a higher salary. He is happy to accept horizontal reassignment to a new job at the same rate of pay, or opportunities to be creative in the job already held. He is motivated to build an identity, to experience control, to have fun, maintain relationships with others, and earn just enough to pay the rent and finance a month of travel overseas—and nothing more. Employers, and older individuals (like me) with very different work ethics are confused: who in their right mind would decline a promotion? But to these young employees, life is not just about navigating to success. It is also about negotiating for experiences that they equate with a life lived well. In this clash of values, there is no fixed solution, just an ever-evolving dance between leaders and followers to decide which sets of values are important.

The rules of navigation and negotiation also apply on larger stages. Recently, I have been making my way through a small library of academic papers on the need for new decentralized energy systems that would foster resilience by putting energy generation closer to energy users. Geoff O'Brien and Alex Hope from Northumbria University have addressed the challenges facing the elderly who live on their own or in substandard nursing homes.[134] In many cases, big centralized energy systems deliver the electrical power those seniors need to survive. The problems of seniors dying from heat stroke when their air conditioners fail, or dying from exposure or carbon monoxide poisoning when their heating systems fail during cold snaps, are easier to solve if we think about localized energy solutions and stop building mammoth outdated energy grids with little capacity to withstand catastrophic weather events. But will we?

In the past, energy resilience often meant centralized coal-fired generators. These days, thanks to technological innovation, it means consumers with solar panels uploading juice to the grid, wind farms, and battery storage facilities either the size of football stadiums or compact enough to hide in a closet. It means decentralized grids and mixed sources of energy for large-scale power generation. It requires dramatic shifts in energy policy and the widespread application of new technology. So, while the solution would help people who are vulnerable—and the list is much longer than seniors—to get the resources they need to cope with life-threatening situations, it would also be disruptive. Progress is always disruptive. Big innovations often threaten a social order, and the processes of navigation and negotiation in these cases are never simple. They reflect competing visions of success, of our requirements for the future.

To be clear, I am not suggesting that we should, or could end our consumption of fossil fuels immediately. In defense of those working in these sectors, and the wealth they generate for us, carbon-intensive industries remain a necessary part of our economy now and will continue to be important in the near future. In communities where my research is taking place, I am hearing stories of new opportunities for employment, education, and a responsible stewardship among industry leaders who are promoting cleaner forms of carbon-based energy extraction and more efficient transmission and consumption. Even in communities affected by climate change, such as the High Arctic, one cannot simply say melting ice is universally bad for everyone. As controversial (even reprehensible) as this may sound, with the opening of the Northwest Passage in Canada's North, more jobs, tourism, and opportunities are arriving that may help solve some of the most serious social problems facing northern communities (while certainly creating many others at the same time).

My point is that change always brings with it some good and some bad. To experience resilience always means trade-offs. But to ignore the advantages of the available improvements is to overlook the additional advantage that disruptions make us stronger and ultimately set us up for more success in the future. Good leadership and good government move the world forward and, with it, everyone who resists change out of self-interest or fear or complacency. It understands that change and innovation are the foundation stones of community resilience. It measures our collective capacity to cope with how well we embrace new technologies and stifle our fear of what comes next. Unfortunately, as O'Brien and Hope write in their study of the elderly, we may not yet be ready or willing to take the necessary steps toward adoption of new patterns of energy consumption, whatever the benefits to our larger society. Our minds and our social processes are proving slow in adapting to these changes, and our institutions are not providing the necessary leadership. We are failing to navigate and negotiate for the resources we need to contend with one of today's most pressing issues.

* * *

Here is a more personal, real-life example of the mirror game in action. I knew a restaurant owner in a small community near to where I live who spent a good deal of time blaming the failure of her business on her staff. The millennials she hired refused to put aside their smartphones, showed up late, and demanded time off for every party and family visit (not to mention Star Wars opening and rock concert). They could seldom think for themselves or take initiative. Eventually, the owner gave up, tired of working 70-hour workweeks supervising bad staff. She could not negotiate with the millennials. To hell with them. To hell with being resilient. She closed the business.

Restaurants fail all the time, and often because of staff problems, but that outcome is not preordained. A family friend of mine who runs a couple of franchises tells a different story of his encounters with millennials. Dave admits young people today will not put away their smartphones, show up late, and can require a lot of supervision and positive reinforcement. Rather than getting upset about this, he has found ways to bring out the best in his employees. Underneath their lack of structure and histories of being indulged as children, Dave has discovered that millennials are a creative, outspoken generation, comfortable meeting new people and taking on new challenges. What an entrepreneur has to do is create institutional practices and supportive environments that bring out the best in young workers. Harsh rules and severe consequences will do nothing except cause trainees to quit.

Dave had one employee, a young man with a bald head, tattoos up his neck, and numerous piercings, who always showed up five minutes late and insisted on keeping his phone with him all day. "He's great with the customers," Dave said. "He's actually quite confident and likes people. Once he's working, he's upbeat, he does anything you ask, and he even tells me about any problems he sees." Those are some of the upsides of millennials. Perhaps they are entitled, but they have also been taught to expect to be heard and to fight for their rights. Some, not all, know how to talk to adults. Many want to find a place to belong and take responsibility for someone or something.

"And the cell phone?" I asked Dave. "How do you get your employees to not be constantly texting?"

"You can't take their cell phones," Dave explained. "That's their lifeline. So I work around it. I tell them they can't be on their phones in front of customers. Ever. But if they have to send a text, they can do that out of sight and on their breaks. They seem to respect that. They might glance at a text, but they know they can't respond. And that seems to calm them down and let them focus. If I took their phone away, I'm pretty sure they'd work less and be even more anxious and distracted."

It is an interesting fix. It fits with what I have learned about resilience being a process of navigation and negotiation for resources in our workplaces, governments, families, and communities. When we structure people's environments to bring out the best in them, most of us do fine so long as we have a say over the resources we get and how those resources (like decision making and access to our cell phones) are provided. Shape environments well, and we can realize remarkable results from people who will be motivated to succeed.

Chapter 8

Social Justice and Success

THE DISTRIBUTION OF RESOURCES in a society always involves value judgments about which parts of our social, built, and natural environments are most important, and to whom. Figuring out who should benefit from the provision of resources is a quagmire of competing priorities. It is also of serious concern among those studying environments that facilitate resilience. While there is no consistent answer as to who should win and who should lose, there is a bias toward solutions that further social justice. At the very least, a social justice orientation seeks to assist the greatest number of potential winners on the understanding that the more winners among us—that is, the more people who experience resilience—the more likely we will all prevail when bad things happen.

A big part of social justice is figuring out how we speak of our experiences, who decides what is important, and what words we use to describe our lives. How do I know that I am being treated fairly if everyone tells me my problems are of my own doing? This is a daily nuisance for people who live with homophobia, sexism, able-body-ism, racism, and other forms of marginalization. Somehow, those in power never fully accept responsibility for their part in creating the messy lives others are forced to live. Fortunately, things do change.

I was at the National Civil Rights Museum in Memphis for the 50th anniversary of the death of Martin Luther King, Jr., who was shot on April 4, 1968, while standing on the balcony of the Lorraine Motel in that city. The motel was converted into the museum in 1991, and its displays were updated in 2014 to better capture how we talk about racism today. Visitors gather first in a small theater where they are given a brief history of the transatlantic slave trade, America's legacy of racism, and African Americans' struggle for freedom and equality. The story is moving, but what impressed me most was that the narrator called the Ku Klux Klan "terrorists."

The use of that one word redefines the narrative of African American experience. It would not have been used five years earlier, but it is entirely appropriate. Racism is a form of terrorism when it is committed by violent, ideological zealots as extreme as any in the Islamic State or North Korea. This is the power of language, demonstrating how important it is to find the right words to capture experience. Until we do so, we are unable to explain to others the profundity of our problems or the feelings of futility we confront when trying to change our world.[135] We are fortunate, these days, that more people are being given the power to speak for themselves. I have heard it said that until the lion gets its own narrator, the hunter will always be the hero of the story. Change language, and the people's ability to describe the injustices they experience changes, too.

★ ★ ★

Kim Jong-un, the boyish anti-American dictator of North Korea, watches NBA basketball. Osama bin Laden watched porn. American pastor Ted Haggard was having extramarital sex with men while preaching against the rights of gays to be part of his church. It is a common story: power breeds entitlement. Dictators and religious leaders live in luxury while telling their people to do without. Everyone needs rules except, it seems, the leaders who impose them.

I worry when I hear about those with power telling those without power that they should be more resilient, that they should work harder, pray harder, or accept life as it is. They are seldom encouraged to fight for fair treatment. The self-help industry is much like these dictators. Rhonda Byrne's *The Secret* tells us that if we think positive thoughts, we will attract good fortune.[136] The sure path to success, she says, is believing in ourselves. She actually believes that we can cure our own cancer with positive thinking. This is the same blame-the-victim, pull-yourself-up-by-your-bootstraps nonsense we have seen before.

If you have read this far, it should not surprise you that equitable access to health care will do more to treat cancer than any amount of positive thinking. Affirmative action to get women into nontraditional workplaces and laws against racism are better sources of resilience than all of Rhonda Byrne's good wishes. Most people, and especially those disadvantaged at birth, those living in refugee camps, and those whose cities have been devastated by war or natural disaster, will only achieve what those in power permit them to. They have more to gain by collective resistance against those who would deny or limit their rights than they do by individualism, presuming they live in a place where their rights are recognized and enforceable.

It is sometimes easier to appreciate the importance of fair treatment when we look across cultures and identify injustices—as if a glimpse of the "other" brings our own world into sharper focus. When I was consulting for the World Bank, I was asked to support a group of Afghani female professors who wanted to understand how to get more women to attend university. Six well-educated women told me in rich detail stories of how girls with dreams of education argued with their fathers, cajoled their mothers, and invoked the support of distant relatives, all for the chance to learn. The barriers they face are incredible. Not only do these young women require their father's permission to attend university, but they also have to find housing and be under the constant supervision of a male family member while attending school. A young woman cannot live on her own in Afghanistan, and universities have very few dormitories where women are welcome or safe. Even if a young woman finds a relative to take her in, and she is not imperiled with the constant threat of sexual violence, her time at university is fraught with difficulties, not the least of which is that some universities do not even provide toilets for their female students.

No amount of self-belief was going to help these young women gain an education. They need social justice, not Bryne's cynical pseudoscience. Although fair treatment comes in many different shapes and sizes, the fundamental problem is always the same. Someone must decide when a right is transgressed, and someone else must be willing to share their power to fix it.

At a global level, the need for social justice is gaining urgency. Frank Heemskerk, executive director of the World Bank Group, said in a recent address that a scenario in which there is no economic growth but where a few people are getting extremely rich will drive people to the streets in protest. That scenario seldom ends well. Egypt, Lebanon, and Hungary are cases in point. So, too, is England after the Brexit vote.[137] Heemskerk is concerned about these protests because major social disruptions are not good for business. They seed uncertainty. They seldom produce enduring solutions to complex problems. Demagogues will take advantage of public anger, saying anything to get elected, even proposing solutions that would ultimately harm the very people they are supposedly empowering.

People do cope better with crises when they are treated fairly. They are less violent and less abusive to themselves and others, they participate more in their communities, and they live more productive lives.[138] In other words, when people are treated fairly, everyone wins, from top to bottom. If we make people more resilient, in the long run, we require fewer jails, emergency rooms, and rehab centers. Once again, the resourced individual is better off than the rugged individual when resources are shared.

In fairness to Byrne, her theory holds some truth if you are a heterosexual, upper-middle-class white person without disabilities, you live in a functioning democracy, and you have a good education and a secure job in a community with a low unemployment rate and affordable housing. It helps if you are male, too. If you are enjoying all of these advantages, then, yes, positive thinking is icing on the cake and likely to predict proportionally more success for you than it would for someone with the same advantages who sees the glass half empty. Most of us are not that fortunate: we will have more luck with shared resources. That does not mean equal resources. No evidence suggests that everyone has to have the same opportunities or wealth. The perfect formula for resilience seems to be that people have enough resources to cope with the stress they encounter.

★ ★ ★

North of the border, it is sometimes said that while Americans were putting a man on the moon, Canadians were giving everyone free health care. Admittedly, free health care is not as sexy as a walk on the moon, but a truly advanced society takes care of its obligations to its most vulnerable citizens before it reaches for the stars. Universal health care, which now exists in every G7 country but the US, strikes most observers as a fundamentally fair system. It allows your child to get access to the same world-class health care as the wealthiest child in your community. You can change jobs at any time without worrying that you might not be able to see a doctor without incurring a massive medical bill. You do not have to put away money in case you break a leg or hurt your back. You do not have to worry about higher premiums because you were unemployed for a year or because you have a pre-existing medical condition.

Decades of research have shown that people who are treated justly do better physically and mentally than those who are not, and we also know that people who are in better health tend to be more productive and happier. I was reminded of this while reading a research brief from *Child Trends*,[139] a US-based organization that monitors child health. It found a connection between access to health insurance and improved child well-being. So, for example, in the United States, 95% of children are covered either by private insurance (52%) or Medicaid, CHIP, and other programs (43%). Those numbers are astounding, first for the fact that one in twenty children still cannot see a doctor for free any time he or she wants, and second because the private insurance business has shut out almost half of all children living in the most industrialized nation on Earth. This is a tragedy.

We know that when children have easy access to high-quality medicine, their mental, physical, emotional, and educational outcomes are vastly improved. Earlier access to health care, usually a privilege reserved only for those with private insurance, means fewer hospitalizations down the road. Studies of American mothers and their children provided with access to an expanded program of Medicaid (on par with health care services in countries such as Canada, Germany, and the UK) showed that the children grew into adults with lower rates of obesity and fewer hospitalizations for a range of disorders. Of particular relevance to mental health, kids with access to good health care also had fewer problems such as eating disorders, risky sexual activity, and drug and alcohol use.

Unmarried, low-income adolescent girls were less likely to become teen parents when enrolled in a health insurance plan.

If this data is still not enough to sway voters to stand up and demand publicly funded health care, consider that children who receive health care are also better learners with higher reading scores and stronger academic achievements. Healthy children also report higher rates of economic security and contribute more to their communities as adults. The evidence is irrefutable. When it comes to health care, there are no alternative facts. The relationship between access to health care and positive outcomes for children is as certain as can be.

My intention is not to idealize any one system of health care: none is perfect. For example, the Canadian system is plagued by long wait times for elective surgeries, including hip and knee replacements.[140] Many Canadians have a problem finding a family physician. There are shortages of specialists in rural areas and the remote North. And, of course, there is the cost: health care is a significant tax burden for Canadians, even though the US system costs more as a percentage of GDP and produces worse outcomes.

Health care, in the developed world, is but one part of a larger social safety net including old age pensions, unemployment insurance, occupational health and safety regulations, child protection laws, and so on. The more generous the programs, the more likely that the most disadvantaged will succeed, and the more people who succeed, the safer, richer, and happier we all are. How many services and what kinds of services people need will always be a matter for negotiation.

In general, social justice advocates tend to follow the principle of universal access to health and social services. Everyone, they reason, should get access to the same services and receive the same tax incentives or family benefits, no matter who they are. This is a good strategy for population health, but it may not provide enough resources to those who are most disrupted by an acute crisis like a forest fire or flood, or when a community is chronically disadvantaged and marginalized like African Nova Scotians in Canada or the poor whites of Appalachia. In cases like these, proportional universality is a better principle for resource sharing.[141] Proponents of proportional universality would target supplementary services to those most at risk. For example, after the birth of a child, many countries provide a year or more of parental benefits so that parents can bond with their newborns. In Canada, parents are eligible for up to 15 weeks of maternity leave followed by 61 weeks of parental leave. While a parent's salary is not fully compensated (it varies between 55% and 33%, depending on the length of time benefits are paid), the benefit does give caregivers the opportunity to take time off work. Even better, mothers and fathers can share the benefit.

Another example of proportional universality is that some jurisdictions offer subsidized child care or breakfast programs in schools that serve mostly disadvantaged populations. Of course, childcare and breakfast are important for all children, but in these instances, governments assume the responsibility to provide both only in at-risk schools. Similarly, school boards might offer smaller class sizes for children from disadvantaged homes, or additional mentors and tutors for students from families with less educated parents. Small class sizes are a great advantage for the marginalized but less important for children from better-resourced homes and communities. Treating people fairly does not mean we all get treated exactly the same or that we all receive the same amount of resources. A difference exists between equality and equity. *Equality* (synonymous with *universality*) means we each have the same right to stand at a fence and watch a ballgame, regardless of our height. *Equity* means that those who are shorter than the rest get a stool so that they can see over the fence. It is the principle of proportional universality in action. Tall people do not need stools; short people do.

All of this has important implications for how governments and other social institutions design and deliver the services that improve our chances of success. A strong case can be made for more equitable disaster relief. After Hurricane Katrina slammed into the Gulf Coast in August 2005, causing 1,833 deaths and $125 billion in damages, we learned a great deal about the rate at which people recover their mental health after natural disasters. Individuals who had stable housing and someone in their household who continued to receive a salary recorded the largest and quickest improvements to their mental health.[142] People with dependent children in their homes recovered less quickly than those without children, indicating an obvious target for relief. People who were forced to live for months, even years, in motel rooms, or suffered long bouts of unemployment, showed higher rates of post-traumatic stress. As we learned earlier from the case of Northern Alberta, the faster we get people back to work and rebuilding their homes, the sooner they can recover their networks, their stability, and their routines, and the faster their lives return to normal. Equitable access to resources will improve a community's mental health and decrease the burden of mental illness on health care systems.

Fair treatment, in our times, means being given the services we need when we need them. I find it odd that one never hears television evangelists extolling the virtues of health and social services or consumer protection agencies. Business gurus are quick with mental exercises to make you a sales superstar but never advise you to request a subsidized gym membership, or paid leave to deal with personal crises or a well-designed harassment policy. After all, prayer is not going to prevent an injury on an unsafe job site, and a good attitude is not a wise coping strategy for sexual harassment on the job. Good things happen because someone somewhere provides a necessary service. Services make us resilient. Government interventions fulfil a social contract that says,

"I accept some rules, I pay my taxes and, in return, I'm not on my own when bad things happen." It says, "My rights are respected. I have the minimum I need to survive: food, shelter, clothing, and safety in my community." This is not wild-eyed socialism. These are practical steps for nurturing success. Anyone who thinks otherwise should try living in a society where the rule of law is fragile and the social safety net is non-existent. Child mortality rates will be high, violence will be rampant, and daily hassles like getting the garbage collected will be unimaginably stressful. The fact is that we are more capable of withstanding shocks when we have systems around us that cushion everyone from exploitation, corruption, and violence.

Two final points: First, fair treatment is almost always the result of group effort. Would we have good labor legislation or laws against domestic violence and child exploitation, would women be able to vote, and would children have the right to an education if people had simply expressed gratitude and expected good things to happen? All of these advances to social policy came about because people refused to adapt to unfair circumstances. The result is that ours is the wealthiest, most secure society that has ever existed in human history.

Second, the social contract is only fulfilled if the services are delivered well. Anytime bureaucratic procedures tie people up in red tape, our communal stress rises and our collective ability to deal with problems goes down. Endless queues when you are renewing your driver's license, endless waits for court proceedings after a divorce or tax refunds or mortgage approvals—these all encourage people to stop trusting each other. We already feel we are in constant competition for scarce resources. Some healthy merit-based competition might build capacity and offer us a welcome bit of structure, but taxing people's patience and endurance does nothing to encourage benevolence toward others.

* * *

Studies of who succeeds and who fails under stress show remarkable consistency over time. Whether we are looking at children growing up in poverty, or employees injured on the job, delinquent young men, or victims of a hurricane, the research shows a 70–20–10 pattern to people's survival rates.[143]

Seventy per cent of people will recover and show minimal impact from their exposure to a potentially traumatizing event *if* they get the resources they require. This is good news: it suggests that most of us our well-resourced individuals will have in place the supports they need long before a crisis occurs. They have a job with a bank of sick days; they have some savings or an extended family who can take them in; they have neighbors or are part of a congregation willing to bring over a casserole, shovel their driveways, or help them care for their children while they are doing whatever they need to do to get through the moment. They live in communities with police, social workers, home care workers, fire departments, ambulances, and food banks. They might have employment insurance, or a pension plan, or a financial advisor to help get them through a layoff. None of this is as sexy as a yoga retreat in Nepal, but our resilience is based on these plodding, boring services and supports that surround us day to day.

A remaining 20% of people, however, will either have insufficient resources to pull through a particularly harsh period of disruption or fail despite having sufficient supports. The trauma they experience will be too much to handle, or their histories of exposure to distress will be too complex and severe to overcome. These people, we know, only succeed when they receive tertiary-level care. They see a family therapist or individual psychologist, get counseling from their clergy, attend court-mandated treatments, or require a period of care inside an institution. They will need professional help finding new work, addictions treatment, or a career coach. Once they are fed, housed, and safe, they may even want to try mindfulness training and other individually oriented treatments, but only after everything else is in place to keep their lives in order.

The dirty little secret that clinicians do not tell clients is that almost all therapies are equally effective when practiced by a competently trained professional.[144] And almost all therapies work for about two-thirds of clients, sometimes more, often less. That means that of the 30% of people needing extra help, two thirds, or 20% of the total population, will respond to treatment, leaving one third, or 10% of everyone in crisis, with enduring problems, despite the care they receive. This is the sad but unavoidable truth about resilience. Not everyone can be saved all the time.

Sometimes, the best we can do is offer people time to change and keep them safe. Grief after the death of a child or the challenges of retirement is eventually resolved for many but for others, there are no easy fixes. Whether because of their genetics, neurobiology, psychology, or social factors, certain people are vulnerable and unable to heal quickly. These are the people who will relapse, find jobs and then lose them, cycle through relationships, and live unhappily ever after. The best we can do as their spouses, employers, and friends is to wait for them to come around. They might even benefit from watching daytime talk shows and reminding themselves that things will get better with time, especially if the world around them shifts a little in their favor.

Time heals because it numbs the trauma and sometimes brings with it the serendipity of new experiences. The sheer randomness of life drops in our laps a new friendship or lover, maybe an opportunity for work or a move to a new community. Sometimes life offers us an inheritance, or maybe the feeling of "dis-ease" that comes from a lack of coherence, both of which can force us to change. Whatever happens, something eventually shifts, and the most damaged among us become ready to engage with a therapist and sincerely work on our problems.

I was thinking about the 70–20–10 rule when I heard about the riots in Ferguson, Missouri. My thoughts went immediately to the people who raise families in communities with poorly funded schools and harsh policing practices that all but guarantee their children will be stopped and searched and possibly arrested, before their 18th birthdays. No wonder they were frustrated and taking to the streets. Missouri is one of the few jurisdictions anywhere among the G20 industrialized countries where school funding is decided at the municipal level, which leads to underfunded schools for those who most need the advantage of a good education. That inequity makes no sense if our goal is to make people more resilient and society more productive and happier. On this issue and others like it, I prefer to avoid the trite ideological fixes to very complex problems promoted by politicians, motivational speakers, and religious zealots. I prefer to be practical and science-driven. I want to raise resilient kids who become resilient adults, who will need less help because they are less vulnerable. To achieve this, we need fair and equitable social structures and good governance as preconditions for success.

If you still need convincing, answer the following 12 questions, which I developed with my colleague, Michele Grossman of Deacon University in Australia, to assess the likelihood that adolescents and young adults will have their resilience undermined by exposure to violent extremism.[145] When answering the questions, think about the community where you live, whether that is your physical neighborhood or an online community with which you interact regularly.

In your community:

1. Are you treated with less courtesy than other   Yes   No
   people?
2. Are you treated with less respect than other people?   Yes   No
3. Do you receive poorer service than other people at   Yes   No
   restaurants and stores?
4. Do people act as if they are afraid of you?   Yes   No
5. Are you called names or insulted?   Yes   No
6. Can people be trusted?   Yes   No
7. Are people willing to help others in the community   Yes   No
   when someone is ill or needs help financially?
8. Do people get along with each other?   Yes   No
9. Do people share the same beliefs about what is   Yes   No
   right and wrong?
10. Would people organize together to keep a fire   Yes   No
    station, clinic, school, or other services open that
    was about to close because of budget cuts?
11. Would people do something about children who   Yes   No
    were skipping school or misbehaving?
12. Would people do something to help if someone   Yes   No
    was being beaten or threatened?

Answers to these questions offer a quick assessment of the likelihood that your community will make its residents resilient, or if
it puts you at a social disadvantage. People living in more resilient
communities tend to answer "No" to questions 1 through 5 and
"Yes" to questions 6 through 12. It is worth underlining here that
the things that make us and our communities resilient are not the
qualities of individuals but aspects of our shared social space. We
could say that our experience of social justice in our community
is a shared responsibility. Whether we are talking about a family,

a worksite, or a neighborhood, as our sense of safety and security increases, our ability to survive the next big setback increases as well. No wonder resilient people tend to cluster together. Find me one individual who is doing well, and I will show you others living and working close by.

* * *

Just as social justice is a foundation stone for resilience so, too, is the basic feeling that we are safe and secure.[146] Like other aspects of resilience, safety and security are not ours to create on our own. They are given to us by our environment, although we can do our part to enhance our experience of them. If we lock away firearms and store ammunition separately, obey traffic laws, and keep our worksites tidy, we keep ourselves out of harm's way and increase our collective capacity to remain safe. In this search for safety, we are both the protagonists who create a culture of safety and the passive beneficiaries of what others do for us. Almost every community has someone who picks up garbage and paints over graffiti. Most of us will call the police when we see a crime being committed. Many people (especially women) benefit from good street lighting at night and by walking together through sketchy neighborhoods. All of these things add up. Efforts we put into our safety and security today are like vaccinations against the inevitable exposure to trauma that we will all experience at least once in our lives. The greater the instability of our living situation, the more that these precautions can help us avoid making our situation worse.

Speaking of vaccinations, they have a lot to teach us about how important safety is to resilience. Parents who refuse to vaccinate kids against diseases like polio and measles are fortunate that they are a minority and that everyone around them is vaccinated. The reason their children can withstand the threat of life-crippling viruses is that they live in communities where the structure of immunization is in place. For some reason, despite all the science, and the outright discrediting of one sketchy study that lied about the connection between vaccines and autism, some parents continue to put their kids in a life-threatening situation. This is an excellent example of how our collective resilience can be undermined by the actions of a selfish few.

We cannot be resilient in contexts of anarchy because no matter how much we believe in the myth of the rugged individual, without some guideposts for our behavior and collective goodwill, our individual actions will never be enough to keep us safe. In other words, we can only act irresponsibly as long as everyone else is looking out for us. The anti-vaxxer is not a danger as long as everyone else vaccinates. But she is a danger to herself and her children if more people start to act irresponsibly. In such a situation, our community changes from being a source of resilience to being the cause of our vulnerability. That is exactly what happened at Disneyland in Anaheim, California in 2015. An unprecedented outbreak of measles among children occurred because the immunization rates had dropped too low to sustain what is called *herd immunity*. In other words, all those kids with the misfortune of being raised by anti-vaxxer parents were suddenly easy prey to a preventable disease because they were all in the same place at the same time.

The same principle of collective safety applies to other more formal institutions than the family. For instance, a group of 13 male dentistry students at Dalhousie University set off a firestorm when they were caught posting misogynistic and homophobic comments on a Facebook page, detailing which of their fellow students they most wanted to rape and fantasizing about how they could sexually assault their patients while they were under anesthetic. The legal options for the university, where I am on faculty, were limited because the Facebook group was private and online. The president of the university responded with a process of "restorative justice." The perpetrators were asked to meet with their victims and be held accountable for the way they undermined everyone's sense of safety on campus. Women felt attacked. Male students who had not participated in the Facebook group felt their reputations were tarnished. Patients at the free clinic operated by the students stopped coming to appointments.

While restorative justice can be a good idea and invoke a sense of belonging to community, it did not bring closure to the abovementioned incident, mostly because it came before the victims (and the rest of the campus) felt safe. When done right, restorative justice gives control back to victims without shaming those who have offended. If it works, victims feel empowered because they have a hand in deciding consequences. Offenders remain in their communities, but rather than simply being punished, they are provided with a roadmap to redemption and restitution. That is what is supposed to happen. In the case of these dentistry students, their continuing presence on campus made many people (women and men) feel unsafe. As a consequence, the process of restorative justice was seen as a way to appease the men (or avoid a lawsuit) rather than a helpful response to the needs of the victims.

Not surprisingly, then, the outcry against the restorative justice process was loud and sustained. If the university had suspended the offenders while the investigation was underway, everyone would have felt secure in his or her classrooms and clinics. With the men still on campus and seeing patients, the press had an easy time finding people who were either furious or afraid, creating a media frenzy that made the situation feel still more dangerous to all involved. Later, in a report that looked at the culture in the Faculty of Dentistry leading up to the Facebook incident, it was found that there had been numerous earlier complaints made against both faculty and students. In other words, the environment was toxic long before the crisis that brought notoriety to the class of 2015. The dean and faculty had failed to provide a safe environment. When a crisis of epic proportions finally occurred, there was little capacity in the school to withstand it, because there had been so little previous effort put into creating a safe environment.

This may sound like it comes from Abraham Maslow's hierarchy of human needs, with the experience of physical and psychological safety being a necessary precursor to the achievement of a sense of belonging, self-esteem, and meaning.[147] While Maslow was not trying to describe the protective processes related to human resilience (his brilliance was providing a theory for human development in general), his work was among the first to link individual cognitive processes to broader social and ecological conditions. The problem I find with Maslow is that he arranged human needs on a ladder as if those at the bottom had to be in place before one could satisfy those at the top. While I agree that safety is important, I am not sure that our other needs can only be met when our safety is assured. In fact, many of the people who have participated in my studies of resilience have told me that their search for a place to belong, a purpose in life, or a powerful identity were all achieved before they ever felt safe. I prefer to think about the factors that make us resilient as the many balls that we need to juggle in the air at once, the position of each ball changing from zenith to nadir in an instant. At any point in time, a particular aspect of resilience can be the key to our success. While I know my university's administration meant well, it made a fatal mistake when it mishandled the competing demands of safety and restorative justice, which is another way of saying that safety should have come before community cohesiveness.

My last two examples have focused on families and postsecondary institutions. Workplaces, too, are just as vulnerable to safety concerns that require creative thinking to keep them just and fair. That is the starting point for Simon Sinek's book *Leaders Eat Last*, the title of which is a line the author borrowed from a sergeant in the American Marine Corps.[148] If the person in charge is going to ask people under her to put their lives at risk (or their jobs on the line), they need to know that their boss is looking out for their welfare before her own. When that social contract is broken, those we supervise can no longer feel safe or treated fairly. Whether it relates to bullying on the job site or sexist and racist attitudes that prevent people from being promoted, ensuring people experience a combination of social justice and psychological and physical safety is always a prerequisite for resilience.

Chapter 9

When Things Work
as They Should

CANADA'S NATIONAL CIRCUS SCHOOL is a sprawling complex in Montreal, across the street from Cirque du Soleil. It was there that I worked with a team of professional circus artists, kinesiologists, sociologists, medical doctors, psychologists, and elementary school physical education teachers on a project to teach children circus skills. Children were given the chance a few times a week during regular gym class to move through a series of stations, each one emphasizing the development of different circus arts. It provided an excellent illustration of how resilience depends on many different systems providing many different resources to maximum effect.

At one station, children climbed aerial silks—long sheets of fabric hanging from the ceiling—learning how to lock their knee around the fabric and hang upside down (they had spotters). At another station, they rode unicycles. At another, they learned to perform tumbling routines and were encouraged to think of all the creative ways they could land their flips and cartwheels. There were stations where they could experiment with juggling, plate spinning, balancing on a roller board, and, if they liked, combining these skills into a single same act.

Standing back against the gymnasium wall, watching the class, I remembered my own gym days, which involved a lot of waiting my turn and intimidation by the kids who were better at sports. Games like dodgeball were hellish for kids of small stature, who were never able to throw the ball as hard as others, and kids with higher body mass indices, who were considered easier targets. Those classes seemed to be designed to get most kids to hate movement, and it worked. Most grew into sedentary teenagers.

Circus provides a new paradigm for teaching physical education. It does this by addressing the multiple biological, psychological, and social systems that need to be activated if we want children to move more. Larger children can be an asset when building pyramids or holding another performer in the air. Anxious children can find a quiet corner to repeat a juggling exercise until they perfect it. Interestingly, we found that children who were talented at conventional sports such as basketball and hockey were least enthused about circus camp. They no longer felt as good about themselves because they were seldom competent at every station they rotated through. After all, the body type and the hand-eye coordination that were advantageous in a full-court press or racing across the blue line were not necessarily helpful in aerial skills or performing a sequence of handsprings and backflips.

The point of this training is more ambitious than simply getting children to move. The exercises develop their confidence to move (a quality of psychological resilience) as well as their physical literacy.[149] That last term may not be familiar to most people, but it is an emerging concept in fields like kinesiology and education. Similar to being able to read and do math, *physical literacy* is about having a set of building blocks for an active life. If I teach a child to play basketball, she will have a limited range of physical skills appropriate to dribbling a ball down a court and shooting baskets. If I teach a child to bounce a ball and run, jump, hop, and leap at the same time, then I have improved the child's capacity to play a number of sports, as well as walk on ice or hike around boulders on a mountain trail. It was exciting for me, as a resilience scientist, to discover that children who participate in circus classes also report a sense of belonging with other children—many circus activities require group cohesion to perform—and better psychological well-being. A 45-minute gym class a few times a week is insufficient to change kids' body mass, but these students also reported feeling healthier and more physically active.

Although definitions of resilience differ across domains of study, scientists are increasingly finding that approaches employing multiple systems, such as circus school, with its biological, psychological, and social dimensions, are especially effective. This is true in disciplines ranging from public health to international development to psychology. There is a growing appreciation for how complex systems work together to improve developmental outcomes, whether that is a child's physical capacity to play or an ecosystem's capacity to withstand an invasive species.

As I showed earlier, resilience, when defined with systems in mind, can be thought of as the ability of one or more systems to anticipate, adapt, and reorganize under conditions of adversity without blowing itself up. In psychological terms, we describe this as *well-being*, but it is a definition that can also be applied to a fertile forest or a profitable business.[150] Success depends on how well each system interacts with other systems. Think of a good movie where the acting, the cinematography, and the score all blend perfectly.

When resilience is understood as a product of mutually dependent systems rather than as the product of an individual, our thinking about problems and their solutions is broadened. Consider our genomes, one of our most basic systems, one we experience without even knowing it is there. It might seem that our genome is a system unto itself, predetermining our fate, but the truth is that our environments, as much as heritability, influence how our genes are expressed. In other words, geneticists are proving that success is as much about what is outside us as what is inside us. You may have a predisposition toward musical talent or exceptional hand-eye coordination, but no single gene or environment is going to make you a rock star or a professional athlete. The story of success implicates dozens of genes being turned on and off by environments at many different levels. Chance events inspire innovation. Our ability to adapt tailors gene expression to the demands of our environment. Genetically speaking, the best we can do is put ourselves into as many novel environments as possible and watch the results. No wonder geneticists remind us that it is our non-shared environments that tend to be the most important to exceptional growth and development. We are all remarkably similar until unique experiences make us different. Once again, environment trumps biology when it comes to resilience.[151]

Having five fingers is a heritable trait. There is nothing unique about possessing all five. Like most of the traits that make us who we are, we share the trait of having five fingers with everyone else. That is because each of us carries a DNA pattern that repeats in every other human. We may enjoy the illusion that we are all originals, but almost all of us share a long list of common characteristics, like five fingers on each hand—which brings us to another paradox. Those same genes that make us the same as everyone else also make each of us different. They interact with systems inside each of us, and with systems external to each of us. And most of the resulting differences are shaped by environment. Biology is not destiny.

Research by Willem Frankenhuis and his colleagues in the Netherlands found that people with histories of child abuse have an uncanny ability to detect threats in their environment, as well as an

enhanced capacity to learn new things and even improved memories for the most relevant parts of their environment.[152] Contrary
to the belief that early trauma impairs *all* functioning, this growing
body of evidence reminds us that human beings are remarkably
adaptable. Needless to say, Frankenhuis's work should not be used
as an excuse to inflict or ignore child abuse, but it gives us hope that
even a bad start in life does not seal our fate: the question is always
one of fit. Do we have what it takes to succeed in the environment
in which we live?

This dynamic, systemic way of thinking about resilience has
made it a popular concept in research about the material sciences: a
coiled metal spring is resilient because it returns to its original shape
after compression. It is also popular in environmental science: a forest that recovers from the insult of an invasive species is sometimes
improved. In fact, resilience studies of this sort are so prevalent that
efforts have been made to chronicle the many uses of the concept.
Jacopo Baggio and his colleagues conducted a citation analysis
to see where the term *resilience* is used, by whom, and where the
term appears to share some common elements.[153] Their conclusion
was that resilience is both a boundary and a bridging concept. As
a boundary concept, it is well-described by each discipline, helping us think about problems in new ways in different domains, but
usually at just one or two systemic levels. As a bridging concept, it
helps us discover patterns of behavior common to multiple systems.
Take, for instance, architecture. An architect who builds resilience
into structures so that they withstand earthquakes is working with
a boundary. An architect who makes a structure that can withstand
earthquakes while also meeting the needs of the people living in it
and respecting the environment is bridging.

This book sees resilience as a bridging concept rather than a
boundary concept. I am curious about questions such as, "How
does a resourceful business make its employees healthier?" or, "How

do people who are loved interact with their natural ecosystems, and how do those ecosystems affect people's attachments in the first place?" Other scientists in different disciplines think about resilience in the same way. Katrina Brown, a professor of social science at Exeter University, studies how international development makes communities more sustainable while also improving the psychological resilience of individuals and the economic resilience of entire economies.[154]

Unfortunately, not enough of us are taking the broad view. Two geographers, Li Xue at the University of Saskatchewan and Yuya Kajikawa of the Tokyo Institute of Technology, undertook a mammoth review of publications on resilience and sorted their findings into 10 disciplines: psychology and social science, business systems and engineering, psychiatry and brain science, marine science and fisheries, and so on.[155] They found an increasing number of publications referring to resilience, but most have remained within the boundaries of one or two disciplines. I was disappointed by the result. We are like the blind zoologists trying to describe an elephant, one touching its trunk and describing a snake, another its leg and describing a tree. Resilience is more than what any one discipline knows it to be. Until we realize that, we will not understand the way multiple systems interact, nor the real reasons we succeed.

Of course, the fact that a small group of scientists are searching for overlapping definitions and instances of cross-fertilization means we are more aware than ever of the need to work together on the mysteries of resilience. This awareness is popping up all over. Leading psychologists and psychiatrists have admitted that their earlier conceptualizations were too narrow. I previously mentioned Ann Masten, who broadened her interests beyond psychological growth after spending time with humanitarian aid workers and ecologists. She now defines resilience as the capacity of dynamic systems to successfully adapt to disturbances that threaten their functioning.

She also holds that resilience as a concept can be applied to any system, human or otherwise.[156]

Shauna BurnSilver, an environmental anthropologist at Arizona State University, has been working with a team of scientists and community elders to explain why some Indigenous communities in Alaska are more successful than others at accessing wild foods and maintaining the health of their people.[157] One might guess it has to do with the quality of the animal herds or the availability of wild plants. The resilience of those ecological systems, however, only explains part of the difference between scarcity and abundance. BurnSilver and her colleagues found that how people shared the food they hunted and gathered best predicted a community's capacity to feed itself. An efficient system of food distribution could offset the precariousness of an ecosystem. As one might expect, the sharing networks had less of an impact when there was plenty of wild food to harvest. In less ideal circumstances, however, sharing compensated for the lack of available resources.[158]

These patterns of adaptation that make individuals and communities more sustainable are what ecologists refer to as *social-ecological systems resilience*.[159] Humans and natural ecologies interact with each other in ways that make each system more robust over a longer term. Of course, ecological systems do not need humans to make them resilient. Many studies have detailed how forests, coral reefs, tidal shelves, and even elephant herds show resilience in their own ways. Natural systems can find many different equilibriums, and whether or not they are desirable to humans is of little concern to social ecologists.[160] Ecosystems do whatever they need to do to survive by reorganizing when confronted with a shock.[161] I am amazed at how social-ecological resilience parallels the processes of psychological and social resilience among humans.

We may have a long way to go, but there is strong momentum behind the idea of understanding resilience as a product of multiple well-functioning systems.[162] Clearly, systems interact. Clearly, we need to cast a critical eye on studies that purport to explain complex human behavior by a single variable. There are no silver bullets, however often we are told that we can improve our memories, our profitability, our community safety, or our academic achievement entirely through individual effort.[163] Resilience can appear in a myriad of forms, with an astounding array of possible behaviors. If we are ever going to understand the many complex systems that affect our success, we will need to become accustomed to taking the broad view of the many systems with which we interact.

We cannot think about a single system at a time. To do so is to build a road for optimal drainage without considering traffic volumes, pedestrian activity, bicycles, emergency services, and communications systems. The road will not flood, but the potential for other problems is high. This is of more than a theoretical concern. James Sterbenz, a professor of electrical engineering and Computer Science at the University of Kansas, looked at several major disruptions experienced by telecommunications centers in the US in recent years, including serious damage to Internet cables in Baltimore after a 2001 fire erupted in the Howard Street tunnel.[164] Many of the companies that required cable connections had chosen to rely on the same tunnel because it provided cheap and easy access to their facilities. That was all they wanted. Traffic engineers, experienced with tunnels, knew that tunnels needed to be built with fires in mind. Confined spaces increase the potential for disaster. When a truck did catch fire in the Howard Street tunnel, its traffic system survived the disruption because engineers had built it with redundancy in mind: car drivers could detour onto other routes, even if not as convenient. This was not the case for the telecommunications carriers: their isolated system was much more vulnerable to disruption. This is as true for humans as it is for the Internet: we do better when we maintain multiple intersecting systems in our lives.

* * *

In the course of my research, I have identified seven principles that make it easier to experience resilience. They enable one to successfully cope with disruptions in the functioning of any system, from a person's own stress response to a competitive business threat. Adherence to these principles predicts successful evolution to new behaviors. In other words, put the principles into practice, and success is likely to follow.

Principle #1: Resilience Only Occurs When There Is Adversity

I had the good fortune to meet Elizabeth Hordge-Freeman at a conference in the mountains of Banff, Alberta. Hordge-Freeman is an African American woman with a beautiful spirit who studies empathy and love and all the good things that make us human. She also digs into the experience of racism and the triggers that make us exclude some but not others. Her work in Brazil is particularly interesting. To better understand why and how people oppress others, Hordge-Freeman interviewed Brazilians who have family members with a significantly different skin color from their own.[165] With Brazil's diverse mix of peoples, a huge potential for variability in skin colors exists. Hordge-Freeman found a close relationship between a child's looks and the love he can expect from a family. Children with darker skin experienced far more adversity at home and beyond the front door—a tragic reminder of the shallowness of racially motivated prejudice.

What I found particularly intriguing about Hordge-Freeman's work was that it exposed an important condition for resilience. It might seem that lighter-skinned children in those Brazilian families showed more "resilience" because they had more success in life; but success is not a characteristic of the children themselves.

Rather, it is the result of growing up in families where their skin color was preferred and they were treated better. It is more accurate to say that darker-skinned children who did well in life showed resilience because they had to overcome more disadvantage than their lighter-skinned peers. For children of African descent in Brazil, their resilience is a barometer of parenting practices and the opportunities that follow. A child with many advantages may claim a little resilience. A child who faces prejudice at home and in her community should be celebrated as all the more resilient if she succeeds.

Without stress, there is no resilience. How systems show their resilience depends on the demands put on them. In the case of humans, we have a range of adaptive strategies, from hyperarousal (becoming very alert) to hypo-arousal (shutting down to avoid becoming overwhelmed). Both are typical ways that individuals deal with atypical levels of stress that activate a body's stress response systems. Of course, when these systems are activated for too long, there is a risk of producing long-term wear and tear.[166] Resilience researchers say a system or, better, a combination of systems shows resilience when bad things happen, but the system(s) manages to find a new normal.[167] This is as true for us as individuals as it is for our economy or the natural environment in which we live.

Let's think about two different kinds of stress on a natural environment. The first is a slow oil leak from a sunken ship that challenges local plants and animals. If the leak is not too large, the ecosystem can adapt fairly easily. At a local level, we would describe this as an example of ecological resilience, as oil spills are not a typical stress for marine life. The surrounding systems—the plants that fish feed on and the fish themselves—remain relatively unchanged.

Contrast that with a huge oil spill that wipes out entire ecosystems—from birds to fish to plant species on the ocean floor. The resilience of all these systems is undermined, and they will need to make major adaptations in how they coexist. In fact, they may die off for a time, until the oil disperses enough to allow micro-plankton to recover and reboot the multiple ecological systems one by one. The new regime will inevitably be different, with new winners and new losers.

Regardless of who or what wins, and who or what loses, observations of natural systems, social systems, and even economic systems remind us that resilience is a process that all systems experience in adversity—hence the phrases that imply positive development when we describe these processes: "urban renewal," "personal recovery," and "economic development." Each implies a change for the better under conditions of stress. Remove the stress and we have "urban planning," "personal reflection," and "community celebration." These are nice things, too, but they are not signs of resilience, and they do not require a science of protective processes to explain how or why they work.

Principle #2: Resilience Is a Process, Not a Trait

For 20 years, the Hadhad family produced fine chocolates from their factory in Damascus, Syria, and shipped their products across the Middle East. Their business ended when their factory was destroyed during the Syrian war and the family fled to Lebanon, remaining there for three years until sponsored by a Canadian church group and brought to the small town of Antigonish, Nova Scotia. There, the Hadhads built a new chocolate business. Wanting to acknowledge the generosity of those who had given them a new home, the company was named Peace by Chocolate. Its tagline is, "One peace at a time." The Hadhad's product has been a hit. The new business was featured in news stories on CNN and the CBC and mentioned at the United Nations in a speech by Canadian Prime Minister Justin Trudeau.

We could say that the Hadhad family's resilience is a result of their individual grit, creativity, faith, personality, or good fortune, but none of these characteristics would have been sufficient without the co-occurring systems that made their success possible. The story of Peace by Chocolate is one that includes an election that brought to power a new Canadian government committed to bringing 25,000 Syrian refugees to Canada; and communities that fundraised $30,000 each to sponsor a family; and bureaucrats that agreed to process claims faster; and economic incentives that helped small businesses get off the ground; and a media industry that likes a good news story now and again. The Hadhad family was welcomed in an abundance of ways. Language classes and educational programs were tailored to the needs of its children.[168] For all these reasons, the resilience of the Hadhad family is the result of a set of processes, which began with escaping a war, that is still ongoing.

As noted earlier, positive experiences of resettlement can be expected with refugees. They are likely to match or exceed economic and social benchmarks of success for the general population of their host countries.[169] Outcomes among refugees as a group are even more impressive if we account for their family structures after they arrive in their host communities. Colleagues of mine who are part of the Child and Youth Refugee Research Coalition, one of the projects I lead, have looked at results for over 700,000 children who arrived in Canada between 1980 and 2000. They found that the children who were settled with a caregiver were most likely to be earning money and filing taxes by the age of 20. This is important as we debate the merits of family reunification or we consider deporting parents even if the children have status to remain in their host countries. Given the clear economic benefit to keeping families intact when resettling immigrants and refugees, we may be imperiling the resilience of entire populations with immense potential.

Whether we are thinking about human biology, psychology, an economic system, or a natural ecology, we misrepresent the science of resilience when we say that a child / economy / forest is inherently resilient. It is not true. Each complex system engages in a process of continual adaptation, responding to its environment and exploiting internal and external resources. Even chocolate makers need to negotiate their way through hundreds of small decisions to reach a successful outcome. Change the resources available and the results will vary.

Principle #3: Gains in One System Come at the Expense of Other Systems

Coral reefs are delicately balanced ecosystems, as interdependent as an economy and just as vulnerable to disruption. The Great Barrier Reef off Australia's eastern coast is currently being destroyed at a cataclysmic rate. Warming oceans, the high phosphorous content of pesticides leeching from farmer's fields, and industrial trawlers that drag nets the size of football fields are all threatening the viability and diversity of the reef ecosystem. Ecologists, tourism operators, and politicians are alarmed at what is happening, despite the fact that some species are thriving in these conditions. As the reef weakens, it is exploited by invasive species of flora and fauna that now find it useful for their own purposes. Large, fleshy seaweeds called macro-algae thrive when the more diverse plant life of a reef is stripped away. That one or two species succeed at the expense of a diverse, well-integrated, and spectacular set of interlocking systems of plants and animals demonstrates how resilient systems are always competing for control of their environments.

Systems eventually reach a new homeostasis after disruption, but the appearance of stability with a new set of behaviors does not mean that every part of a system, or another related system, will be happy with the new normal. Trade-offs will always occur when systems experience resilience, with some parts of a system benefiting more than others. Furthermore, systems not struggling with problems are generally weaker than systems competing for scarce resources to survive. As discussed earlier, competition and stress exposure produce a steeling effect by which disturbed systems become better able to cope in the future.[170]

Catherine Panter-Brick and Mark Eggerman at Yale University have studied the rigid family structures of Afghanis and shown that high expectations and constrained choices have protected families against external threats through centuries of war in Afghanistan.[171] Family members know their place, and their social order can withstand the assault of changing political structures. That same rigidity, however, causes individuals to be frustrated with their lack of choices, and that frustration can turn into mental illness. While we think of resilience as generally benefiting multiple systems, there are always winners and losers under each different resilience strategy.

Likewise, a farm foreclosure may be tragic for the people involved but advantageous for ecological systems that thrive when fields lie fallow for a time. The trade-off may be difficult to see due to our bias toward the needs of humans over those of plants and animals.[172] We prefer to think in terms of maximum sustainable yields and high commodity prices, forgetting that the cultivation of monocrops like corn, soybeans, and wheat exact a toll on an ecosystem's diversity and sustainability. We reserve the word *resilience* for systems that we perceive as improved but forget that one system's loss is often another's gain.

Principle #4: Open, Dynamic, and Complex Systems Show More Resilience

Equilibrium? Homeostasis? Stability? A resilient system foregoes these concepts and prefers to be disturbed, perturbed, and mildly threatened. A system exposed to manageable amounts of stress is open as new resources become available and accessible. These resources are what can help systems remain complex enough to handle bigger crises when they occur.

An effective and well-managed health care system, for instance, must develop contingency plans for future shocks by incrementally training itself to handle small but immediate crises. Staff routinely practice drills so that every subsystem, from intake to the intensive care unit, knows what to do should the unthinkable happen and the emergency room receives hundreds of casualties at once. We might also say that health care systems are more resilient when they participate in knowledge sharing and implementation. Even well-trained physicians take over a decade to adopt best practices that can save lives—a marked improvement from centuries ago, when surgeons refused to believe that the blood on their gowns was actually posing a risk to their patients. Of course, a certain amount of skepticism benefits patients: quackery is everywhere, as is pseudo-science, promising to cure childhood leukemia with vitamins.

Stability may feel good in the moment, but long term, it is a lousy way to experience resilience. Americans have almost completely stemmed the tide of refugees at their borders and turned a blind eye to resettlement needs of displaced populations globally. While I can understand the fear, it is nevertheless misplaced. Many of the first settlers from Europe were fleeing persecution, and they did all right (although the same cannot be said of the Indigenous peoples they displaced). Whether we focus on the success of the refugees who arrived after the world wars, the Vietnamese "boat people" of the 1970s, or the recent waves of Iranians, Guatemalans, and North Africans, displaced populations tend to carry with them strategies for survival and talents suited to a vibrant economy. With a near-zero unemployment rate today in the United States and a figure close to that in Canada, there is little reason for people living in G7 countries to think that refugees take jobs from native-born citizens. Again, there is every reason to expect that in as little as a decade, refugees will perform as well or better than long-term residents of the host countries in economic terms.[173]

Principle #5: A Resilient System Is Connected to Other Systems

Resilient systems tend to encourage heterogeneity, but it is never enough to say, "My business is diverse." Resilient systems also need to promote connections between the different parts of a system, and between one system and another. The more collaborative the network—the more businesses work together to advocate for their industry—the more likely the network will succeed.

Think about how human and natural systems interact. Fishers are often fighting among themselves on issues of sustainability: one group wants a particular species in a particular habit to be protected; another group wants a different species in a different habitat protected. Örjan Bodin's studies of fisheries, governance structures, and sustainability of both people and fish stocks have demonstrated that a sustainable fishery is not focused on just one part of the ecosystem.[174] Rather, it builds a network of fishers who collaborate so that no one species is overharvested and no one habitat is unintentionally destroyed. A fishing technique such as dragging nets along the ocean floor, which may increase an individual fisher's yield, has been banned because it can destroy the habitats required for species like lobster, upon which other parts of the fishing industry depend.

As Bodin explains, overharvesting one species can threaten supplies of food and nutrients throughout the ecosphere. Ecologists thus talk about fish stocks as intricate webs of mutual dependency.[175] The individual excesses of a single boat can exact a terrible toll on other parts of the industry. This is not a matter of winners and losers: the exploiter will realize only short-term gains because the destruction of co-occurring systems means that the entire ecosystem and the human communities that rely upon it are also imperiled.

All connected systems are vulnerable to negative influence if a single part fails. A single murder or kidnapping in a resort community can cause bookings to dry up for every hotel in the area, even if they had nothing to do with the violence. A single hotelier might run advertisements boasting of his thick stone walls and locked wrought-iron gate, but his establishment will be smeared along with the others. The only way the community will recover is if a network of hoteliers and other businesses and civic leaders work hard to re-establish the community as a holiday paradise. Systems that work together thrive together.

Principle #6: Resilient Systems Like to Learn and Experiment

Nothing like failure sharpens one's wits and gets the creative juices flowing. Artists accept failure as a familiar bedfellow, teasing it awake until they find the inspiration they need. They are examples of resilient systems that embrace experimentation as a key to their success. They have a penchant for risk-taking, or at least a willingness to push themselves into what the Russian psychologist Lev Vygotsky referred to as the *zone of proximal development*: that uncomfortable space between what you know you can do and what you will never be capable of doing.[176]

Connectivity shapes opportunities for experimentation, while supportive environments give us solid foundations from which we can take chances. Systems that are more innovative will inevitably show great stamina. The Catholic Church, for example, has survived a series of existential threats following the Protestant Reformation of the 16th century by constantly adapting itself to its changing environment. In recent times, Pope John XXIII convened the Second Vatican Council to make the Church more ecumenical and accessible, often using vernacular languages instead of Latin for its mass. Another set of changes came with Pope Francis, who has liberalized Catholic doctrine on a range of issues from divorce to climate change.

Ideally, feedback between artists and their audiences, priests and their congregations, or any other related systems will push people to experiment with new solutions to persistent problems.[177] Resilient systems do this well. Every change brings with it the possibility of new opportunities for success.

Principle #7: A Resilient System Is a Rich System

Systems that are open increase their diversity, which makes them more successful. The more complex a system is, the more pieces that participate in it, and the more likely it is to succeed. If it is complex, a system is also likely to have redundancies, backup plans for when parts fail. These three things—diversity, participation, and redundancy—make for a rich resilient system.

Diversity brings systems sufficient resources to anticipate future stress. The more parts of a system that are engaged in ensuring its welfare, the more likely it will remain viable. Extended families are great for raising children, especially very young children who tend to need a disproportionately large number of human and financial resources. Businesses thrive when it gives opportunities to employees to participate in some of the decisions that get made—for instance, brainstorming solutions to workflow processes. The more diverse a system is, the more the system will succeed at solving complex problems.

Participation matters: the higher the rates of community involvement, the healthier the community. Jeffrey Liew at Texas A&M University conducted a three-year study of a multiethnic cohort of six-year-olds to track their academic achievement. He found that the abilities of the kids were only part of the story.[178] The classroom environment, the quality of peer interactions, the relationships between students and teachers, and the competence of the educators all made substantial contributions. The takeaway is that a bunch of systems are required to bring out the best in a student. The more each system or part of a system participates and helps a child succeed, the more likely that child is to do well. This is even more likely to be true when a child is socially marginalized by race, ethnicity, or language.

Systems often fail because they are not complicated enough to withstand the stress to which they are exposed. Political dictatorships may look resilient when they resist change for a time, but they lack long-term resilience because they have low participation and no redundancy. Topple a Saddam Hussein or Muammar Gaddafi and factionalism and civil war are more likely to occur than a smooth transition. Redundancy is crucial. Every time I am on a plane, I am reassured by the fact that there are two engines, each capable of keeping the plane aloft should something catastrophic happen.

The trick with redundancy is to have enough to keep all systems safe but not so much that a system's efficiency is compromised. Too many redundancies when risk is low will make a system clunky and inefficient. As we have seen, overprotective parenting in safe communities contributes to a spike in anxiety disorders among children. Excessive regulation on resource extraction industries has all but shuttered them in some developed nations. And too many regulations on trade in the European Union is giving right-leaning demagogues the fuel they need to stoke nationalism.

When we get redundancy right, along with diversity and participation, the results are typically strong, rich systems with the capacity to withstand progressively heavier burdens. The three characteristics support one another. Diversity, by its welcoming nature, encourages participation. Diversity and participation encourage redundancy: someone or something is around to pick up the slack. Resilience supports resilience—that is one of the wonders of the concept. These principles appear in every resilient system that I know, whether it is man-made, natural, or human.

Chapter 10

Strategies for Resilience

AMONG THE MANY MYTHS I hear about resilience are that great jobs and feeling loved are good for everyone in equal measure. I am certainly not against these resources (in fact, I quite like having a healthy dose of each in my life), but what is good for one person may not suit another if his past experiences convince him otherwise. If you associate well-paying employment with angry bosses who belittle you or being loved with manipulation and sexual violence, then it makes good sense to avoid both options and go looking for other resources that feel more congruent with how you experience the world. This might mean remaining on social assistance and living below the poverty line or choosing to rescue cats from the humane society rather than get into complicated relationships with two-legged creatures. These patterns of coping make sense when I put people's choices into context.

Unfortunately, and despite the need to think about context before thinking about solutions, people still share what they believe are elixirs for lifelong happiness without any thought to whether they will work for different people with very different lived experiences. I see this often at house parties when someone drops the news that they have been laid off or their spouse is leaving them. The amount of advice people give seems to be in direct proportion to the number of empty beer and wine bottles lined up on the kitchen counter. "Get a better job," "Find yourself a better lover," or, more crudely, "Get laid (or drunk)" are the pat solutions that (even sober) people offer to solve other people's very complex problems. Even though we may wish that a one-size-fits-all strategy for success could overcome every challenge, the truth is that as stress increases, the strategies that help us succeed change, too.

Studies of resilience and the many interventions that research has seeded tell me that when I am in crisis, I can (1) change myself, (2) make the best use possible of the opportunities I have so that change happens, (3) change the world so I have more opportunities to be my best, or, when all else fails, (4) change what I want so that I am less disappointed when nothing changes at all. Which theory of change is best will depend on how much risk I face day to day. I have seen time and time again that when people experience higher levels of risk exposure, changing individuals one at a time is not going to create long-term personal transformation. At lower levels of risk, however, it does, as long as the world keeps dishing out resources and providing the opportunities we need to exploit them.

Coping Strategy 1: Change Myself

Every episode of *Scared Straight*, a psychologically abusive bit of reality television about juvenile delinquents experiencing jail, makes the assumption that we are self-made individuals capable of shaping our destinies. If we do not like where our lives are now, we can do something about it. We can try harder, find new friends, make smarter and better choices, and become our best selves. The good news is that this life-changing strategy works, but only when the stress we face is manageable and we have the personal resources to excel. If we are realistic about people's opportunities to realize their potential, exposing a juvenile delinquent who is unlikely to re-offend to the terror of prison life should be a deterrent. He thinks, "I don't want to be here. I can do better than this." And he will have the opportunities to change. But a high-risk kid entering jail will think, "I can handle this," or "Now I'm certified tough." He will see jail as a rite of passage, an experience on par with the biggest roller coaster at the amusement park. Delinquency is a great coping strategy when you have few other options to access the resources you need to succeed.

While some people in some privileged circumstances can change themselves, it only works to a point. The reality show *American Idol* shows us the winners who have had enough support, coaching, and opportunity to grasp the brass ring. What they never show are the auditoriums full of talented people who were model students, who believed in themselves, prayed, and remained positive day after day, yet could not get beyond the first round of competition. It would be harsh to blame these individuals, or God, or their teachers, or their parents for their failure. They did everything they could to improve their skills and exploit their talents, but they did not appeal to the particular judges on this one show. They might shine in other circumstances. Sometimes it is our environment that lets us down.

A young man I met in one of my studies applied to medical school three times and was rejected each time. To his credit, he never stopped believing in himself, which was also sad, because he could have done so much with his life if he had dropped the illusion of unlimited possibilities. In the competitive world of academics, he was not a contender. The barriers to becoming a physician were high and his resources few. He was a kind and caring individual who might have made a great physiotherapist, paramedic, or licensed care worker. I think he might have preferred a profession other than medicine. After all, the things he valued most about helping people were unlikely to occur in the five-minute-per-patient world of a family practice. This young man *should* have changed himself. He had options, and he had enough money to enroll in any professional program he wanted. It made no sense for him to remain focused on failing at medical school. A better path to resilience would have been to choose a different career path. That he did not is the kind of thing that happens when we believe our inner worlds are more powerful than the world we experience around us.

Coping Strategy 2: Make the Most of Opportunities

Maybe you are a married woman in your late forties dealing with demands at work, with children moving out on their own, and an elderly parent who is refusing to move into a senior's residence despite a couple of recent falls and the early signs of dementia. That is a lot of stress to manage with positive thinking. Or maybe you are early in your career and unable to find an entry-level job where you live because foreign-trade deals have made it easier to move blue-collar jobs overseas. Perhaps you are retiring and realizing that your days are empty. Changing yourself is not going to work. We cannot think our way out of caring for an elderly parent, or out of unemployment, or out of a lonely retirement. We cannot re-create ourselves overnight and expect everything to be okay. When we experience acute stress (like an injury or job loss) or chronic stress (responsibility for an invalid parent), we need a more effective way of coping than standing in front of a mirror and saying, "I can do it!" The truth is, most of us cannot, and will not until we look for new opportunities and the support we need to manage our stress.

As stress piles up, the best coping strategy is to maximize resources.[179] We are more likely to succeed when we reach out across our social network for help or take advantage of opportunities when they present themselves. Even better, we can make opportunities happen. There is a story of a man sitting at home when the police come to his door and advise him to evacuate because of a flood. The police offer to transport him to a shelter, but the man politely refuses, telling the officers, "God will save me." When the flood comes, the man has to climb to the second floor of his home to keep from drowning. A boat comes by to rescue him, but once again he waves the help away, saying, "God will save me." The waters continue to rise, and now the man is forced onto his roof. A helicopter lowers a line, but, again, the old man refuses to be rescued, insisting, "God will save me." He arrives at the gates of heaven to meet God, with whom he is visibly disappointed: "Lord, you didn't save me!" God says, "I sent you a truck, a boat and a helicopter. What more did you want!"

Self-reliance and the illusion of personal control are likely to cause us harm when the odds are stacked against us. Our salvation is in the opportunities in our environment: a new job or a change in responsibilities; relocating the family or retraining to get more marketable skills. Retirees join service clubs, volunteer, work part-time, or develop hobbies. Adults with elderly parents can turn to financial counselors to learn how best to use their parents' savings or assets to fund their care. It is rare that we have no opportunities at all. Often, we only need help in identifying them.

Coping Strategy 3: Change My World

What if you cannot change jobs, or fund the care your elderly parent needs, or take on more student loans to retrain? What can one person do against insurmountable odds? At some point, we need to change the world around us if we are to deal with greater and greater amounts of stress. One person can actually do quite a lot against insurmountable odds if she stops taking responsibility for things she cannot change and asks for help to change the things she can.

The adult with responsibilities for an elderly mother who cannot afford a senior's residence can benefit from a social worker to help explore other options. The young person with few employment prospects can benefit from a school-to-work co-op placement partially funded by business or government. A stay-at-home parent missing adult conversation can benefit from an informal meet-up group at the local Starbucks. If supports like these do not exist, create them. If we need resources that are not being provided, advocate for them. Call your union (that is what it is there for) or your local politician (that is what she is there for). Contact the United Way or another local service agency. Ask for what you need from whoever has the resources to help.

Most of us can instigate small changes to the world around us if we are motivated to do so. This might simply be ordering a new desk to avoid carpal tunnel syndrome and lower back pain. It could be requesting a microwave in the lunchroom. It could be asking city council to put a speed bump on your street to slow traffic. Maybe we want to petition for lower tuition fees for students or protest against police profiling. Maybe there is a political party we can join or an occupational health and safety issue that we can champion in our workplace. From the ridiculously local to the impossibly global, we are more resilient when we experience the sense of personal and social efficacy that comes with meaningful involvement in changing the world around us.

Whether it works or not, what is clear is that changing the environment can matter more to an employee's quality of life than anything an employee can change on his own. The same goes at home. Often the best solution to a problem is a change of environment. More than one family I know with a delinquent child has sold its home and settled in a new community so that the child will find a different peer group. This is a dramatic example of changing environments, but it works because it shows the child that a new start is possible. Despite our best efforts to change, living in a place that reinforces old habits makes it difficult to become a better person.

Coping Strategy 4: Change What You Want

Bear with me for a moment as I discount most of what I have written in the past nine and a half chapters. Despite all the things we can do to change our environment and be more successful, sometimes the walls around us will be too high, too strong, or too many to climb. We will be weighted down with a large number of problems, and no matter how many allies we have, how many opportunities we exploit, or how much we try to change the world around us, we will continue to fail. In that situation, whether of your own making or a conspiracy of ill fortune, you can keep trying, or you can accept your defeat. At moments like this, the surest way to cope is to change what you want.

Refugees stuck for years in a UNHCR camp would go crazy if they expected their lives to change quickly. Sometimes, it is better to give up the dream of returning home, or resettlement, and make peace with the world around. Even in refugee camps, parents find ways to educate their children, secure more food, and make themselves safe. They fortify their walls and do what they can to survive. It is far from ideal, but it is a strategy for resilience nonetheless.

Our workplaces can be just as obstinate. What do we do when we hit the glass ceiling and advancement is no longer possible? We can look for new opportunities; we can suggest to our employer ways to improve the business and our own situation. Yet despite these efforts, many people remain stuck in jobs for many years doing the same work for the same pay. They cannot quit because of financial or family obligations. So they wait. They cope by drinking or saving for an early retirement. Sometimes they have affairs, or they play endless video games. They pay down their mortgage or put their kid into after-school tutoring to make the daily sacrifice seem worthwhile. They do anything to make themselves feel better about the future. These are all excellent strategies, but they do not change the individual's experience at work.

In these types of situations, we may have to forget the promotion and focus instead on being good at something elsewhere in our life. We have to stop thinking about earning more money and think instead about attending that one concert we always wanted to see, for example. We can find small victories close at hand, ignoring the bigger struggle we have already lost. With time, grief and anger are massaged away and we adjust to our circumstances—at least until circumstances change: our boss retires; a war ends; we inherit money. In an instance, new opportunities emerge, and we can reconsider the other three coping strategies. I am all for hope, but hope in the face of complete and undeniable disaster leads nowhere but to depression. Sometimes, all we can do is change expectations and wait until the world around us changes for the better.

* * *

It might seem like we have four *choices* of coping strategies, but the moment we say *choice*, we imply that our actions are under our control. This is not true. Just because we believe we can change the world does not mean the world will change. Many of us live with the fiction that with enough willpower, we can change any personal habit or flaw and surmount any obstacle. The dominant narrative of our time, the one we see portrayed in Hollywood and Bollywood movies, is of exceptional individuals achieving more than anyone expected them to achieve. I am not saying this is impossible; I am saying, as a researcher, that it is unlikely. For every *Slumdog Millionaire*, billions of others toil in abject poverty wishing only for their next meal. Our choices are limited by the harsh realities of our social and physical environments. The worse the environment, the fewer doors are available to open. To fail to acknowledge this is to blame victims of misfortune for their circumstances.

None of these four strategies is a one-time choice, and none is exclusive of the others. The best coping strategy is always the one easiest to implement and best tailored to the risks we face in a particular situation. In my clinical experience, those who succeed tend to use a series of strategies. First, they try to be their best. They read self-help books, explore their potential, and do as much as they can on their own. When they reach the limits of that strategy, they look for resources and opportunities within their lives to support their dreams. If they find them, their search is over. If they do not, a more likely outcome, they summon the time, motivation, and passion to change the world around them in ways small and large. Only when these strategies fail do resilient individuals abandon their dreams and accept the things they cannot change. This is not permanent resignation: it is just a temporary adaptation to the injustices of an unfair world.

To illustrate how people move through these coping strategies, consider a single mother of three who wants to return to school and train for a better job. She has many stresses and few resources to make such a dramatic change. She has told the people around her of her intentions and insisted to doubters that she has the talent to do better in life. Besides, she reasons, it will be good to model the value of a postsecondary education to her kids. This last argument resonates with people. By recasting her mission as a way of benefiting her children, people begin to accept her vision of what her life could be. Research on success affirms that a mother's level of education is one of the best predictors of a child's academic success. She has used the first coping strategy to see herself as a college student and muster the confidence to tell others of her dream.[180]

Unfortunately, she still does not have the resources to return to school, and because no amount of positive thinking will fill her bank account, our mother moves to the second coping strategy: exploiting the opportunities that are already present. She learns of opportunities for scholarships and loans. Maybe her workplace has an employee assistance program that will pay for individual coaching. Maybe a relative will volunteer to look after her kids while she goes to class, cutting her daycare costs considerably. The list goes on. The world is not always rich with opportunities but seldom is it a wasteland.

Making progress with her plan, but not enough to clear the obstacles ahead of her, our mother adopts the third strategy and begins to consider ways of changing the world around her. Can her expenses be trimmed by renting a smaller apartment while she is in school? Does the college she wants to attend offer subsidized daycare? Can she ask her current employer for a year's leave of absence? Can her parents be materially supportive of her dreams? Or can they at least stop calling her a "bad mother" for trying to make time to study? Can she find new friends who will admire her attempt to better her life? None of these strategies challenges the status quo or storms the barricades of power, but they do subtly challenge assumptions of what single mothers can do with their lives. The more voices that speak out for change, the more commonplace change becomes, and the better resourced we are likely to be.

Step four? If all else fails and the world does nothing to help her retrain, then, and only then, will she have to accept that nothing will get better quickly. At that point, she will have to tell herself, "I tried," and focus her attention on getting her children into college instead of herself. Maybe her dream was too audacious and too badly resourced for the moment. Shelving a dream and imagining a more modest one does not make great reality television, nor is it what we want to hear from those who are supposed to inspire us. I have yet to see a daytime talk show promote "give up and move on" as a coping strategy, although every one of us at some point in our lives takes this path to resilience.

Having worked with hundreds of families over my career, I am convinced that honest effort with the first three coping strategies will usually make the fourth unnecessary. The world is far readier to help us realize our goals than we think. The required actions are often quite simple. The shopping addict who can see that easy credit is her problem will find that her bank will gladly lower her credit limit—a small change in her environment that requires little motivation on her part. A person suffering from loneliness can work an hour or two a day at Starbucks, with its free Wi-Fi and comfortable chairs, removing some of the isolation he feels while working at his business. Whether we make changes through our individual actions or are lucky enough to live next door to opportunity, the world will usually accommodate us.

★ ★ ★

People cluster into two groups when it comes to resilience. There are those with tragically few opportunities to change their lives but who still manage to find resources. And then there are the fortunate. These people have the resources they need right in front of them—all they require is a push in the right direction to overcome their barriers. It is often assumed that people with the resources they need to succeed will choose the saintliest of coping strategies. We assume that the well-educated third-generation immigrant will acculturate; that our depressed co-worker with the high-paying job will find meaning in his life; or that a death in the family will help everyone feel closer to God. I can understand our desire to make saints out of those who overcome adversity, but you would be surprised at how often I meet people who choose less inspiring expressions of resilience. There are many unconventional paths to success, and the choices of resilient people are not always pretty.

Sometimes the best way to handle a speed bump is not to slow down and pass over it: sometimes we need to slide onto the shoulder and steer around it; sometimes we say "to hell with it," and accelerate, launching ourselves into the air, regardless of the danger to ourselves, our vehicle, and others. Not every strategy people use to get through their days is socially desirable. Some are downright funny. Others would start revolutions if we all used them. Resilience is like that. Many paths exist, to many different ends, all of them experienced as successful by those whose way forward is blocked. Here are some of my favorite unconventional strategies for success gleaned from research and clinical practice.

Let Yourself Fail

Tim Brown, CEO and president of the design firm IDEO, reminds us that those who fail early succeed sooner.[181] He gives the example of the inventors of a scale to weigh newborns. Getting a child to remain still long enough for an accurate weight measurement is difficult. The design team thought soothing music might help, so its first prototype included speakers. Surprisingly, the music startled children rather than soothing them, prompting them to cry rather than doze off. The entire idea might have been a failure except that the team found that babies startle, they also stop moving. The music did work, just not in the way intended.

Learning and growing, experiencing creativity, requires us to be uncomfortable, to risk failure. We need to get into that space where we feel mildly stressed but still competent enough to muddle our way through problems. Our resilience depends on our ability to navigate outside our comfort zones. Every individual I have met who describes herself as resilient has told me stories of being forced to work through challenges. It is unfortunate, then, that so many of us avoid discomfort. We go to Disney World or an all-inclusive resort rather than venture into places unknown. We routinize our lives to the point of obsession and nag when our schedule is breached. One of the sad features of our time is that so many people are protected from the stresses that would help them grow.

The opportunity to fail should be welcomed rather than shunned. Failure is the basis for all human development— it presents new perspectives on old problems. In research, we call these *negative findings*, and they are usually the inspiration for groundbreaking ideas. Penicillin was discovered because a petri dish became moldy. A railroad construction accident in 1848 that put a lead pipe through the skull of Phineas Gage provided the catalyst for a theory linking personality to neurological structure. How many of us have jobs we like because of a wrong turn and a chance encounter, or because we messed up an interview with the firm we thought we wanted to work for? When we have the resources to fix our mistakes, failure, even tragedy, can be a powerful advantage.

Let Yourself Be Stressed

The humorist David Roche has a vascular disease that caused a serious disfigurement to his face at birth. Growing up, he was teased and excluded because of his looks, but he also learned to cope with difficult situations.[182] Among his strategies was a commitment to court stress. He spoke in public and became an advocate for social change. He took his political campaigns on transit buses, shocking passengers with his impassioned words while enduring insults for both his looks and messages. Every time he was put down, he used the experience to grow stronger. He is a remarkable individual who also benefited from a hugely supportive family.

There is great value in pursuing stress. When my clients' lives have become too predictable and they have lost their zip, I suggest introducing more stress into their routines rather than less. "Go meet someone new, try a new sport, plan a new career," I tell them. In his book *Adapt*, Tim Harford says we need to tolerate, even promote, failure and change if we are to experience long-term success. Just as failure inspires us, so too can stress keep us sharp.[183]

When you think about your past, is it not true that many of your positive experiences occurred when you were at the edge of your comfort zone? When did you earn your first dollar, and how did it feel? When did you move out on your own? When did you first have sex? Each of these new, stressful experiences brings with it potentially huge payoffs (I am assuming the sex was consensual and satisfying). Resilience flourishes when we encounter stress in manageable amounts. While most of us try to avoid it, resilient people seek it out. So never mind clearing your mind in an ashram. Fill your life with activities, tackle life and its challenges, and enjoy the sense of forward motion all your activity brings.

Be Soft Rather than Hard

Some books about resilience have titles like *Grit*[184] and *Mindset*.[185] They preach perseverance and individual strengthen. We tend to applaud people of perseverance and strength, but we rarely celebrate those who wiggle their way around problems, avoiding confrontations, and still succeeding. Sometimes the best way to solve a problem is to approach it in a softer, less direct manner.

Harold Warner, the founder of Dynamic Air Shelters, an engineering firm that builds blast proof buildings for military and industrial applications, has a lot to teach about soft approaches. His shelters must withstand an explosion caused by either an industrial accident or an act of war. Conventional wisdom says to build shelters strong, we must make their walls thicker and sturdier. Warner, however, used his experience as a competitive ballooning champion to invent canvas shelters that withstand blasts better than cement and steel structures. The principle is remarkably simple: a cement wall must absorb a blast, and its fragments frequently kill the very people they are intended to protect. Canvas walls, cleverly engineered, absorb a blast by redirecting energy away from the structure's occupants. Watching a video of a dummy sitting on a chair inside one of Warner's buildings while a bomb goes off outside makes my brain hiccup. The walls buckle, but the air inside remains so calm the dummy remains sitting.

Sometimes, the most strategically resilient thing is to do less.

There are moments when an admission of weakness in front of others is the best way to absorb a threat to our well-being. If you are assigned another task with another tight deadline at work and feel that you cannot handle it, that you are being set up for failure, give back some control over the situation to the person who is delegating the tasks. Say to your boss, "This new task looks very important, but I'm not sure I have enough experience to know which of the other tasks on my desk at the moment I should put aside. Can you help me decide which has the highest priority?" Admittedly, this might not make one look like the best employee, but returning responsibility for future failure to the supervisor can be a remarkably effective strategy. If the employer cannot determine which tasks need to be done in which order, how can the underling?

This is not passive-aggressive behavior: it is a strategic request for fair treatment. By momentarily looking weaker than you are, you gain a measure of control overt he situation. Sometimes, the softest people, like the softest buildings, are the ones most likely to succeed.

Get Less Help

My friend has a 90-year-old father with leukemia. He experienced a slight spike in his temperature on Thanksgiving Day, and she took him to the hospital immediately. The elderly man waited 15 hours before doctors determined that his condition was stable. In the meantime, he had endured a sleepless night on a bed in the emergency room and the risk of infection from the other patients. More treatment is not necessarily better.

In most instances, we are more resilient when we resist the temptation to overwhelm ourselves with help from others. We have built a society that equates more services with better health when, in fact, getting the right amount of service helps us the most.

As the sociologist John McKnight observed decades ago, we have replaced the rituals of community support with the services of professionals.[186] It used to be the case that after the death of a spouse, parent, or child, one's neighbors and friends would come by to console the grieving. The vast majority of people need only the continuity of friendship in such moments, but we are hesitant to act this way today. A visit to the therapist's couch is now required to return to well-being after the death of a loved one.

The same principle applies to anxious children when their reluctance to attend school is indulged by bubble-wrapping parents. When it comes to treating a child's anxiety, I reference Lynn Lyons, an expert on children's mood disorders based in Concord, New Hampshire.[187] She recommends acknowledging a child's dread but reintegrating him into normal activities as quickly as possible, even if he feels uncomfortable. She also helps children find a safe place at school to go when overwhelmed but the message to parents is clear: their child's anxiety should not be overly pathologized or overtreated. Too much attention on what is often a normal phase of child development can create long-term psychopathology. When it comes to resilience, less intervention is often a more effective strategy for success.

Be Selfish, but Not Too Selfish

We all need to embrace our inner bastard. It is to our advantage to shed our illusions of saintliness and accept that our survival sometimes depends on looking after ourselves. While Buddhists talk about selflessness, business gurus encourage vested self-interest. The truth is that people who are resilient understand that their personal motivation for wealth and status is protective, even while their success requires good relationships with others. It is not much of a surprise that the right amount of selfishness makes us more resilient. Individual greed can motivate creativity. Selfishness can spur initiative. Never underestimate the power of doing what is in your own best interest, and do not be afraid to act first and ask permission later. You can always ask for forgiveness after you have succeeded.

I have witnessed the power of self-interest in places as distant as Botswana and Namibia, where the solution to tackling the problem of big game poaching has been to ensure jobs in the tourism industry for people in the communities that house poachers. Self-interested communities are themselves the most effective way to police illegal activities that threaten local livelihoods. The power of greed, when tempered by social consciousness, can make businesses and entire communities more viable. Western capitalism may have problems, but it has proven to be the engine for the greatest advances in economic and physical well-being in the history of humankind. I would be hard-pressed to identify more than a handful of transformative technologies that came from religious theocracies, totalitarian regimes, or anarchic states.

Self-Disrupt

Some of my colleagues and acquaintances confuse their mortgages with a prison sentence. The huge difference, of course, is that you can get out of a mortgage any time you like with only a small penalty. Yet, for some strange reason, people get tied down to things like houses and sofas when what they actually want are experiences—travel and time to play with their kids. They buy a large home in the suburbs and get locked into high-paying jobs to cover the expenses for the second car they need to commute. If that scenario describes you, try calculating the amount of money you spend each year on your car. Include depreciation, interest on the loan, insurance, license, repairs, and fuel. Then calculate how much you need to earn before taxes to pay for that car. It is not uncommon for people to estimate car expenses to be at least $6,000 a year, which means having to earn anywhere from $8,000 to $12,000 before tax. Now ask yourself if there is a job you would like more that would pay less but not require the commute or the second car.

People who are resilient know how to self-disrupt. They recognize when life sucks and do something about it. I am not saying to quit your job today unless you are confident you can land an alternative. I am not saying to sell your house at a huge loss. I am saying that people who show resilience know when it is time for a change. If they want to have a child, they stop waiting until conditions are perfect (I have five children and can swear that there is no perfect time). If they feel their job is in peril and they know they should relocate, they relocate early enough to take advantage of the best new opportunities. If they have a passion for travel, they do not wait for retirement. They negotiate leaves without pay as early in their careers as they can.

Self-disruption is also good for communities. When the military was shuttering unneeded bases in Canada, one of the first to close was in Summerside, Prince Edward Island. Politicians and local business owners forecasted dire consequences. The opposite happened. The closure prompted the community to diversify its economy and turn the old base into an aeronautics industrial park with aircraft maintenance facilities and a training facility developed by a local college. Summerside benefited by being the first base to close. If it had been the last, many of these opportunities would have been snapped up by other communities.

Be Future Oriented

Spiritual gurus tell us to live in the moment; and when I meet people who are resilient and coping with difficult situations, they sometimes tell me they are living in the present. Nevertheless, they always seem to have a plan for their future.

I was thinking about such things while swimming in an infinity pool on the rooftop terrace of the Morrissey Hotel in Jakarta, Indonesia. The pool is a long blue splice between the teak decking and the dusty chaos of one of the world's most stressful cities. Your attention is drawn to a sky crisscrossed by building cranes, neon billboards, electrical and telephone lines, washing hung out on rooftop clotheslines, and minarets of mosques ringed with speakers to call worshippers to prayer. In an infinity pool, you are never fully present, because you are always watching the horizon for what comes next. To experience resilience is much the same. We have to appreciate where we are, but we also have to anticipate the dangers in front of us and the solutions we have at hand. Resilience is in equal parts being here and being there. Having goals makes us resilient if only because the act of anticipating the future motivates us to push ourselves to seek something better. The spiritually inclined may beg to differ: they speak of selflessness and having no expectations. I prefer to put my belief in saving for my retirement and hoping compound interest and the NASDAQ composite index do what they are supposed to do.

Forget Romance; Find Someone to Whom You Matter

The acclaimed couples' therapist Sue Johnson[188] likes to share advice from her mother: "Love is a funny five minutes, so never trust a man." She may not be warm or scholarly, but Johnson's mother, who owns an English pub with her husband, reminds us that the intense feeling of love is usually brief. We sometimes think that resilient people have earth-shaking love affairs or perfect bonds with loving spouses. From talking with many successful individuals, I can say with certainty that they, too, argue at home, lose it with their children, and divorce individuals they once considered soul mates. What makes them more resilient than most of us is the depth of their social capital. They know they matter to someone, even if that person is not their lover. It may be an elderly parent who relies on them to visit, a young child who needs them to be strong, a friend who leans on them when bad things happen, or maybe even simply a pet that demands affection. True vulnerability is screaming into the night with no one to listen.

Esther Perel, another well-known family therapist and author of the international bestseller *Mating in Captivity*,[189] told me the story of a woman whom she was counseling whose husband worked long hours and earned a sizable income. Whenever she asked for her husband's time, he pointed to the luxuries that surrounded them and called his wife ungrateful for forgetting that it was his long hours that provided these things. Perel's client, however, did not want more things—no diamonds, in-ground pools, or dinner parties. All she wanted was 20 minutes a day of her husband's attention, just wanted to feel like she mattered. This is a simple feeling, and much easier to find than true love.

Make Money

Maybe it is obvious, but it is nearly impossible to be resilient without the means to support yourself. Making money is the key. Match your lifestyle to your earnings. Avoid debt, especially those payday loans with exorbitant interest rates. If you can earn enough to save a little, even better. Developing a marketable skill set and doing what you can to maintain a minimal cash flow will do more for your sense of well-being than any brand of therapy. Yes, resilience is related to income. Spending money on trivial items or accumulating debt will decrease resilience. A steady salary is an advantage only if it is not squandered.

Be Patient

Some species of pitcher plants close for self-protection during periods of drought. Adolescence eventually ends, and awkward teens find like-minded peer groups in their twenties. The pain of separation after a divorce eventually subsides as new routines establish themselves. Even the most abusive parent eventually passes away, and that wicked boss of yours will eventually get promoted out of your life. When all else fails, resilient systems and people show an uncanny ability to be patient. They sense that the forces that hurt or threaten them today will one day be gone from their lives.

While this strategy is partly a mind game, it is nevertheless practical. Your enemies will eventually get old, fail, retire, or die, and when they do, your life will get better. If you are lucky, the spouse you have not liked for years may finally have an affair and give you the opening you need to get a divorce. Resilience is part patience and part optimism: it expects that eventually our lives will get better.

Be Gender Flexible

King penguins share responsibility with their mates for hatching eggs. I find that amusing and inspiring, especially when I think about the Philippines, a traditionally patriarchal society. There, men have adapted to their wives working overseas as domestic laborers.

Fathers routinely assume most of the household chores, including childcare, while their wives are away. Military families in North America show a similar pattern, only in reverse. While the military is diversifying its ranks, more men than women are still deployed overseas. When they go, their wives mow the lawn and get the car repaired, tasks that are more typically done by men.[190] Stereotypes notwithstanding, both individuals and families are more resilient when they adhere less strictly to gendered identities. Shrugging off strict gender norms creates more opportunities for success and better access to resilience resources.

Shock Your Neighbors

Day to day, there is no better way to assert a powerful identity than to shock your neighbors, family members, and colleagues. Try it. I guarantee you will feel stronger and more in control of your life the next time something bad happens. Start small. Let the dandelions grow on your front lawn for a week or two and call them flowers. Put them in a vase on your dining room table. Replace your regular desk with a standing desk and watch how people notice you (especially if you are now able to look out over the walls of your cubicle). Travel somewhere exotic and brag about it to your family. Go skinny dipping, if that feels like a risk. Wear a bow tie (unless you're a hipster, in which case, don't). Don't shave your face, legs, armpits—the choice is yours. All of these small acts of defiance assert who you are. They tell others that you live outside the box, or at least want to. You may find your actions contagious. Others may suddenly start performing their own acts of defiance. Applaud them and I promise they will applaud you back.

You would also do well to surround yourself with deviants. Most of us live in social spaces that prefer conformity over individual control and happiness. It is difficult to experience resilience on your own.

Find other deviants, individuals who like to live outside the box. You are stronger through association and mutual reinforcement.

Avoid Evil Do-Gooders

Being resilient means avoiding those who exploit our need to feel strong. It is easy to see how Islamic fascists exploited disenfranchised youth to become foreign fighters in Syria, but the same dynamic is at play elsewhere. Self-promoting Christian preachers scam money from their audiences. Some G20 activists want nothing but an opportunity to burn a police car, to rebel for the sake of rebellion. Some psychotherapists use their clients to make themselves feel omnipotent. There are environmental zealots who refuse to consider reasonable compromises, even though they drive cars to protest rallies. Falling in with these kinds of people does not enhance our resilience. It only leaves us more vulnerable.

Give Yourself Roots and Wings

Resilient people are both anchored to place and liberated enough to seek opportunities. I have encountered cultures where families encourage a child to emigrate yet bury her placenta outside the front door as a permanent reminder of where her home is. Having a sense of place, and of home, is useful when disaster strikes.

There are dozens of other unique strategies for resilience. Altogether, they leave me astounded at the extent of human ingenuity under stress. What all the strategies share is an emphasis on the world around us and how that world must be engaged and changed if we are to experience individual success.

Chapter 11
Beyond Self-Help

THERE ARE LIBRARIES OF ADVICE on how to stay active, get fit, eat well, and trim your waistline. Right beside them are all the books on how to extend your longevity, manage your time and your budget, and live an organized life. These are the staples of the self-help industry, although it is not at all clear that they help anyone but the authors and publishers. Most of the advice is faddish pseudoscience, if not outright quackery. We are supposed to believe that blueberries, rippling abs, real estate investments, clean closets, and the habits of French women are the keys to a successful life.

The gurus behind all of this advice usually turn out to have feet of clay. They are seldom the best advertisements for their ideas, as Pagan Kennedy of the *New York Times* learned from reading obituaries of the founders of fad diets. Roy Walford, who argued that a strict calorie-reduced diet could double the lifespan of mice, died at the ripe old age of 79 from Lou Gehrig's disease. Robert Atkins, whose high-fat, high-protein diet gave cardiologists nightmares, suffered cardiac arrest in 2002. He later fell on an icy sidewalk and died at the age of 72. Kennedy developed a long list of diet gods who turned out to be remarkably mortal. His conclusion: "It's the decisions that we make as a collective that matter more than any choice we make on our own."[191]

Just as our environment shapes our mental health, it also predicts our physical well-being. Once again, nurture trumps nature.
The social, political, and natural environments in which we live
are far more important to our health, fitness, finances, and time
management than our individual thoughts, feelings, or behaviors.
If proof is necessary, consider Dean Kriellaars of the University of
Manitoba and an experiment he did with children.[192] In an effort to
increase their physical activity, he and his graduate students used
accelerometers to gather baseline data on the number of steps
and the intensity of children's movements in a large elementary
school. They next went into the school on a weekend and painted
hopscotch squares in the middle of the hallways. The results were
inspiring and entertaining. As children moved between classrooms,
Kriellaars was able to document a significant increase in the children's activity levels. They jumped more and stepped more, loving
the opportunity to combine play with an everyday task like walking. A small change in the children's environment changed their
behavior without requiring any change in motivation.

Motivation to change should not be the deciding factor when
it comes to long-term health. We know that the vast majority of
diets fail and, worse, result in changes to metabolism that increase
people's weight afterwards.[193] Gyms rely on the people who buy
memberships in January to not clog the change rooms come March.
The self-help industry thrives on the false promise that emphasizing
"self" solves your problems. An ecological understanding of success
does not discount individual motivation entirely. It simply places
the focus on changing the environment to create the conditions
for healthier living. This is the logic behind putting calorie counts
on menus: it makes people aware of the consequences of their
orders. We may hate being told what to do by our governments,
but our health depends on top-down initiatives. Regulatory efforts
to wean us off our cigarettes by pairing anti-smoking campaigns

with changes to the availability, marketing, and cost of the product paid a huge public health dividend.

The same approach does not work for every problem. Campaigns to promote higher activity levels have largely failed.[194] All they have managed to do is change our attitudes toward being active. We feel guilty for driving to the store rather than walking but grab our keys regardless. To have an impact on how many steps we take each day will require building cities with more mass transit and fewer highways, denser housing stock and less suburban sprawl.[195] In the absence of those policy initiatives, we will have to rely on ourselves to stay healthy, and the data tells us we are failing miserably.

The best approach to staying fit is to live somewhere that demands we move more and eat less. Urban planners can do just as much for our health as diet doctors and fitness coaches if we let them.[196] Where there are sidewalks, people tend to walk more, especially if those walkways border water or a natural landscape. Even public transit has the added advantage of increasing the number of steps we take each day, getting to and from a train or bus. By far, the unhealthiest places to live are North American suburbs. They breed dependency on cars and provide no convenient way to walk or bike to services. That is not the case in many parts of the world, especially Europe, where transit systems and a culture that promotes cycling and walking have translated into better health outcomes.

I was recently in Sweden, which has to be the healthiest place on Earth. It was late autumn and I remarked to my host, Mehdi Ghazinour, a criminologist, on how busy the bike paths were even as the temperature approached freezing. "In the winter," I asked, "do people take buses or use their own cars?" Mehdi smiled. "The town plows the bike paths and we bike," he said. "Why would we do anything different?"

The Swedes like to say there is no such thing as bad weather, just bad clothing. Proper winter jackets that are affordable and made from fabrics that both insulate and breathe are predictive of a more active lifestyle. Good bike paths also translate into more active people. And, yet, we continue to build communities that depend on cars, with the grocery shops several miles away. It may sound counterintuitive, but living in an urban core can be healthier than the distant suburbs if we look at the amount people move and their access to healthy food.

At the risk of oversimplifying a complicated science, physical health is available to most citizens, even in the worst of environments. It is more difficult to achieve, however, if others create spaces that discourage healthfulness. This does not mean that we need to attach treadmills and bicycles to desks or jump on small trampolines at our lunch hour to increase activity levels during the day. These tactics get points for novelty, but we would all be better off if social norms encouraged better physical health. Staircases in hotels and office buildings are difficult-to-find cement tombs, forcing us to take the elevator instead. The ubiquity of the chair makes it mandatory to sit rather than stand at public meetings and business conferences. And would it not help to have a decent place to park your bike and shower at work? Building fitness into our lifestyles and changing our environments to habituate us to exercise will always do more to promote health than strategies dependent on trips to the gym for an artificial period of exercise. Regimes that require atypical patterns of behavior depend too much on individual motivation.[197]

How much movement is enough to promote resilience? An optimal level of fitness is about being able to do the things we want and need to do. Play with grandchildren, hike a short distance on vacation, put down mulch in the garden, or stay on our feet all day at work. We succeed if we train our bodies to be in constant motion. Biking at one's desk is unlikely to become a norm, but standing desks are a reasonable compromise that help workers avoid problems associated with a sedentary lifestyle. I am dubious of claims that we need to run hard for an hour a day or follow any cardiovascular regime that will give us the heart and lungs of an Olympic athlete. Recent research indicates that what we need most are bursts of activity to bring our heart rates up, even if they are relatively short and infrequent.[198]

★ ★ ★

Healthy eating is mostly common sense, but we fail to restrain ourselves every time there is a birthday party or a staff luncheon. We eat too many calories, and they compromise our physiological and psychological functioning. It is no surprise that we eat too much. Our brains are hardwired to consume extra calories when they are available in the expectation that lean times will follow. Our ancestors lived under constant threat of famine. A good diet has to override our instinct to eat by reminding us to do so only when we are hungry.

Deprivation diets do the exact opposite to us. They cause us to stress over food, contributing to the likelihood of increased calorie consumption, causing weight gain in the long term. During the Muslim month of fasting called Ramadan, devotees avoid any food or drink between sunrise and sunset. Once the fast is broken, the tendency is to eat far more calories than necessary. Christians who abstain from indulgences during Lent often ramp up their consumption once the period of devotion has ended. Perceiving deprivation as a danger, our bodies overcompensate by urging us to eat whatever and whenever we can. While I can understand fasts as an expression of personal devotion, from a physiological perspective, they are not great for long-term health.

One strategy for breaking this cycle of starvation and binging is called *eating awareness*, which encourages us to monitor hunger and cravings rather than caloric intake.[199] Before eating, ask yourself how hungry you are on a scale from zero (full) to five (ravenous), and try to discern what your body really wants to eat. Keeping a logbook of responses creates an external record that helps condition eaters to feel satisfied rather than bloated. Add to this a fridge stocked with foods that meet our nutritional needs (rather than our reckless appetites), and our bodies will direct us toward a healthy diet and, eventually, an ideal weight.

My 103-year-old father-in-law Wallace grew up on a farm and to this day eats mostly wholesome foods he grows in his garden. His diet is bland, by my standards—boiled dinners with little seasoning—although I will admit that my mother-in-law's fresh fruit pies and homemade bread make up for the routine of the five standard vegetables found in their freezer. Like many people who live long, the food Wallace eats is unprocessed and readily available from his environment. For millennia, before J. B. MacKinnon and Alisa Smith invented the 100-mile diet, humans matched their nutritional needs to their environments.[200] They ate more protein when doing physical labor, and more fat and carbohydrates when it was cold. They followed the seasons and the demands of nature. Longevity and a healthy diet depend on people adapting to their environments and taking what they needed nutritionally rather than on strenuous and usually short-lived exercises in self-restraint.

Another problem with fad diets is that they pay little attention to the external factors that trigger weight gain. When our environment, including our pantry and fridge, is stocked with the healthy food that our body needs, there is less opportunity to reach for potato chips. Overeating is not just a problem of willpower; it is a problem of food availability and the practices of a food industry that make it easy to eat poorly.

Dozens of studies of childhood obesity show that environment counts more than individual choices. Researchers have monitored changes in the body mass indices (BMI) of students after their school cafeterias dropped burgers, fries, and other unhealthy fare from the menu. Most of these efforts failed to make kids lose weight.[201] They would just wait until they got home to load up on empty calories. The only programs that had a positive impact on BMI were those that also worked with parents to change the foods available at home. Solving a problem like childhood obesity will require the systemic approaches discussed in earlier chapters.

The way we use our time, and the support we receive for our style of time management, can also dramatically influence our success. Sleep is a poignant example of this. Arianna Huffington, the founder of the *Huffington Post*, wrote in her best-selling book *The Sleep Revolution* that she was proud of her ability to function on less than four hours of sleep a night until she flamed out and realized the value of sleep.[202] Most of us need between seven and eight hours of sleep per day, but few of us get it. No surprise, then, that we often need to catch up on sleep on weekends and while on vacation. Optimal functioning is achieved when we get the right number of hours for our brains to recharge. Suffice it to say, resilient individuals know when to go to bed, and those blessed with helpful environments are able to get the sleep they need. Quiet bedrooms with dark curtains and a comfortable mattress help, as does a spouse or nanny to tend to fussing children so that we can keep our eyes closed a little longer.

Our waking hours are just as important. People who are successful seem to know how to balance work and play. Work is easy: it is structured and provides immediate payoffs in the form of financial rewards. It is the play sides of our lives that are difficult. Resilient individuals take time to laugh: they sing in choirs, kick a ball with the kids, or play in adult ways. Sex is playtime, too. So is tending a garden or renovating the basement. Some adults participate in organized sports. Many of us hike, bike, or swim for recreation. The activity does not matter as long as we find it satisfying. All of these activities are, of course, easier to engage in when our communities facilitate access. For example, community gardens using reclaimed urban land have become a much-valued resource, especially where access to recreation and healthy food is in short supply.

As with eating and movement, our use of time will always be circumscribed by those around us and the degree of personal empowerment we experience. I prefer approaches to time management that avoid artificial scheduling of what to do and when. It is far better, it seems, to follow our own natural rhythm or match our rhythm to the task at hand. Good time management is not about being forced into 30-minute slots but rather finding the right time to give our best to the task at hand.

For decades, I have used the early mornings before my meetings with staff at the Resilience Research Centre for writing. I write best in the early hours. Afternoons, as per my natural rhythm, I reserve for more social activities, such as clinical work with clients. I leave mindless tasks such as answering emails and (with sincere apologies to my students) grading undergraduate essays until later in the evening when I am less creative and more easily distracted. I settled on this approach to time management after reading a wonderful book, *Tao of Time*, by Diana Hunt and Pam Hait.[203] There are few greater luxuries than having control over when one does the tasks one needs to do. Arriving at 7:00 a.m. and leaving by 3:00 p.m. can be a life-changing experience for workers with young families and those who have to fight their way through congested commuter routes. Likewise, where there is sufficient trust and accountability built into the structure of the workplace, working a day or two a week from home can increase productivity for some employees. The *Tao of Time* works, but it requires a facilitative environment.

★ ★ ★

Financial success is just as much a quagmire of competing claims as the advice we get about our physical health and time management. Should we save 10% of our income or a specific amount each month? Should we pay down our mortgage or invest in a bull market? Is renting really better than buying? Does getting a university education ensure a lifelong return on the investment? The best answer I can find to each of these questions is, "It depends." As environments change, we must also change.

My partner, Paula, spent almost 20 years as a financial counselor for the military. Many soldiers encounter money and relationship problems after they return home from overseas deployments, and a financial counselor helps return a family to a measure of stability. This stability gives a soldier peace of mind enabling him or her to better resist the aftereffects of war. Oddly, at the same time that the Canadian military is increasing its spending on psychologists, it is stripping away the services of financial counselors and other support personnel. What is the point of making soldiers feel better on the inside if they return home to unpaid bank loans and a mortgage three months in arrears? Financial resilience is just as critical to well-being as having the mental and physical capacities to cope with stress. Too often, we overlook this mundane aspect of people's lives and the way finances and household debt promote or undermine psychological and social functioning.

This was the case for a young man who came to see Paula the year before she took early retirement. She listened carefully as the soldier described his position as a seaman and told her about the many nights he had spent at bars and the "big boy" toys he had bought: all-terrain vehicles, a pickup truck, big-screen televisions. Payday loans had stopped the bleeding momentarily, only to leave him with huge unpaid bills and a surging pile of compounding interest charges. Both his electricity and his cable were about to be turned off after months of nonpayment. Paula is not one to tell people what to do, so when the soldier asked her which bill he should pay first, she shrugged and smiled, assuming he knew. He paid for his cable.

Financial sustainability is still more crucial to mental health and social well-being when we experience adversity. Financial resilience means having enough cash and fixed assets to cope when bad things happen. While there are many things we can do individually, financial health is shaped by the world around us far more than we think. The Great Recession that began with the housing crash of 2008 was the direct result of the banking industry in the United States acting irresponsibly and giving people mortgages they had little hope of meeting. The entire economy turned into one giant pyramid scheme while bankers took huge bonuses. What many Americans may not know is that not every country suffered the same fate. In Canada, a more regulated banking industry made subprime mortgages, 40-year unsecured mortgages, and other exploitive practices illegal. The Canadian housing market, while far from perfect, did not crash, nor did any of its banks.

Say what one will about government regulation: if applied well, it can support the resilience of an entire nation. We are all more resilient when surrounded by solid financial institutions. We have little choice but to trust them; few of us understand enough about finance to navigate banking and credit systems on our own.

Talking about money is never easy, especially with family. Financial resilience nevertheless often depends on having those tough conversations sooner rather than later. Much as we might like to avoid the thought, much of our life comes back to simple economics, even after death. A dear friend of mine has elderly parents with considerable wealth, though you would never know it from looking at their 60-year-old home with its original melamine countertops. They have at times been generous to their children, giving them each a loan of $50,000, which they expected to be paid back. Five of the six children complied, but one has left the debt unpaid. Whenever the topic of money comes up at family gatherings, you can tell they are all looking at him, but no one has dared to ask when he will pay back what he owes.

Families are constantly tying themselves up with financial arrangements. In this instance, the elderly parents may not need the money, but it will be a serious point of discord after their funerals. The children will be seized with the inequity of one not paying his debt. It will be difficult to say exactly how much money was loaned and not repaid, and there will be questions about whether those funds should earn interest. If there is to be an equal division of assets, who will decide what is fair? Will anyone have the stomach to deal with all this and grief at the same time?

Financial health is about planning and keeping it simple. It is about having the touch conversations and structuring our finances to avoid conflict or exposure to exploitation. That can mean a prenuptial agreement, a well-crafted will, savings for a child's education, a simple budget, and adequate disability and life insurance. It is also about insisting on proper government regulations to protect us. None of this is as sexy as offshore tax shelters or cannabis stocks, but it will deliver a far more sustainable experience of happiness when the inevitable occurs.

Being financially resilient means being able to bounce back from an unexpected money crunch, but it also means spending enough to enjoy one's quality of life before one is too old to enjoy it. My clients often report greater happiness when they knock at least one item, maybe even two, off their bucket lists before they retire. Taking advantage of opportunities (and creating opportunities when they do not exist) can change the way we are perceived by others, creating a virtuous feedback loop in which we feel better about ourselves and experience greater fulfillment in life. If, for you, that is reaching Machu Picchu in the Peruvian mountains, then save and go.

How much is the right amount to spend? There is no perfect formula. Even David Chilton, better known as the Wealthy Barber, who used to advise putting aside 10% of one's income into long-term savings, now says that is unlikely to be enough.[204] Times change, and so do market conditions and interest rates on fixed-term investments. Compound interest is no longer the reliable friend it once was. These days, we have to think about tax efficiency, investments in property, and whether or not our children will take responsibility for us in our old age—or at least place us in a nursing home where our care attendant changes the bedding now and again. All the same, a good financial plan will help us withstand the stress of changes to our health and future circumstances, and as is so often the case, our environment will likely do more than our investment counselors to facilitate our resilience.

One final thought about money: Our problem as a society is not that we all lack cash. It is that we all lack time to enjoy it. The vast majority of Americans do not take all of their annual leave.[205] This is ridiculous and, worse, unproductive. Unfortunately, their workplaces could care less. If our environment does not provide us with the support we need for rejuvenation and enjoyment, we are more likely to go under when life turns rough, no matter how much we've read about swimming with sharks.

Chapter 12

Our Endless Potential

I AM ALWAYS ON THE LOOKOUT for stories that make us feel optimistic about our capacity to thrive, counter-narratives to the many mean-spirited attempts to punish the vulnerable or blame them for conditions beyond their control. If you follow the media, you can be excused for believing we live in a world awash with heartlessness and xenophobia. We have been stunned by Brexit, the rise of the alt-right in the United States, and fascists in Europe. We were shocked to see disadvantaged people losing access to health care and families being ripped apart by harsh immigration policies. Life can seem bleak. Yet, at the same time, we live in an age in which beer companies, of all organizations, are staking their brands on the marketability of such concepts as cooperation and mutual respect.

A famous Budweiser commercial that aired during the 2017 Super Bowl told the story of a young German immigrant who, on arriving in America, experienced all manner of prejudice, bullying, and danger before going on to create the iconic American brand, Budweiser. Its rival, Heineken, followed with a five-minute video called "Open Your World," which pit people of strongly divergent political views against one another to prove there is more that unites us than divides us. These advertisements are a testament to the true values of most people. The beer companies appreciate that most of us want to see ourselves as tolerant, liberal, and open-minded, regardless of the content of our Twitter streams.

It is fascinating to find brewers staking out this ground because mass advertisers understand better than most the power of environment. They study public attitudes and susceptibilities to ensure their messages will hit home; spend fabulous amounts of money to create the right images, the right conditions, to maneuver us into seeing things their way; and carefully choose the contexts—which television programs, which websites, which magazines—their advertisements appear in. And we all accept advertising as part of our environments and recognize its incredible persuasive power, its ability to shapes our attitudes, needs, and desires. Yet even when these masters of environment are lecturing us in their advertisements on the importance of the world around us, we still tend to think our resilience is dependent on our New Year's resolutions, our individual motivation, and the thoughts in our heads.[206]

Regardless of the fact that it is delivered in the service of beer sales, the message of these advertisements is worth heeding. Tolerance is a prerequisite for our collective capacity to thrive. Our world is becoming a more globalized, integrated whole, and there is no stopping these trends. The more we connect and intertwine our economies, the more we break down travel barriers, and the more we share our cultures, the stronger, more flexible and more resilient humanity will become. The future depends on our capacity to be good to ourselves and others, and I have witnessed enough resilience in individuals and communities to be confident in our capacity to meet that challenge.

Another organization showing a surprising interest in the role of environment on resilience is the National Aeronautics and Space

Administration (NASA). To succeed as an astronaut, it turns out, one requires a particular assortment of personal resources, groomed over time in environments tailored to the specific needs of space travelers. Individual qualities of grit and daring celebrated as "the right stuff" in the early days of the space program, are not enough. NASA is now preparing to send humans to Mars. To do so requires a deeper understanding of the dynamic relationship between people and their environments.

Psychologists at the University of Nebraska Lincoln were tasked by NASA with instilling in astronauts the mental toughness required to survive a multi-year mission. NASA expects it will take six months to travel to the planet, and that the astronauts will need to remain there for 18 months until Mars and Earth perfectly realign for the return voyage. That adds up to two and a half years in a metal can of one form or another. For most of that time, the astronauts will be too far from Earth to be able to see their planet (something no humans have yet experienced). Fred Luthans, a management expert with expertise in psychological capital, has developed a profile of candidates most likely to survive the experience, and methods to train people to excel in those circumstances.[207] His is a task as critical to mission success as the technological breakthroughs required for sustaining life in space.

Previous attempts to simulate a Mars mission and find the right kind of astronauts have not gone well. The Mars500 simulation back in 2010–2011 sequestered six people for 520 days to see what happens to confined people under conditions of numbing inactivity punctuated by extreme crises. All of them experienced psychological or physical problems, ranging from sleep difficulties to extreme agitation. The actual voyage will cause unfathomable stress, requiring the astronauts to draw on all manner of internal and external resilience resources. We are still a long way from knowing what those resilience resources should be.

What I find most interesting about this research is how flexible NASA has become with its definition of *resilience*. Age and life experience have become more important than youthful exuberance. Technical expertise is still important, but so too are social traits. Months of isolation demand astronauts have personal characteristics that most of us neither possess nor want. NASA prefers people who have a low need for affection, something we only associate with resilience in the most traumatic of circumstances, including war and domestic violence. Conscientiousness and extroversion, traits prized in terrestrial workplaces, are less desirable when your world is shrunk to the size of a garage, and are potentially harmful in an extreme environment where perfection is unlikely. NASA is saying that the personality traits that facilitate resilience are always contextual.

This has been the principal message of this book. Environment matters. Relationships with other people give us the psychological building blocks we need for self-esteem and personal agency. Schools, communities, and even our government shape our opportunities to experience well-being, making our goals more or less possible depending on whether they fit with the priorities of those in power. Housing, transportation systems, and other infrastructure can either limit our access to the resources we need or facilitate access to the experiences that will transform our futures.

Elizabeth Yee, vice-president for City Solutions with the Rockefeller Foundation's 100 Resilient Cities project, has been developing ways to re-engineer urban space to enhance human well-being. Paris, for example, after successive heat waves, realized that it needed to address the problem of urban heat islands that were killing people, especially elderly who were subsisting on limited pensions without access to air conditioning and in social isolation. Working with city planners, Paris's chief resilience officer pointed the way to a simple, effective solution. Paris has over 700 schools, most with outdoor yards and big trees. Most people in the city live within 200 meters of one of these schools. Here was a gift of physical infrastructure that happened not to be in use most of the day. Efforts were made to turn those schoolyards into green oases that drew people out of their homes and into shady areas where— an additional benefit—social services could find them. Paris's green solution to a human problem is more common than we think. A tree outside a hospital window can decrease a patient's use of self-administered pain medication; children who play outside are more likely to be attentive when back in the classroom.[208]

Once again, our success and health depend as much on our surroundings as on our thoughts and feelings. No matter what your genome has given you to work with, there is likely someplace on Earth (or Mars) where the constellation of personality traits that make you who you are can be put to good use. The fidgety, hyperalert child with a diagnosis of attention deficit disorder might be a teacher's nightmare in an otherwise calm classroom but set him in a dangerous, fast-moving environment and he will be the child most likely to prevail. When you find your place, you flourish. When you do not, you are forced to adapt in ways that discourage your unique contribution to the overall good of society. My advice: rather than assuming that you have to meditate, cleanse, or change, look for a niche where you can thrive as you are. With a surprisingly small amount of effort, we can find better places for ourselves and optimize our environments to bring out our best. Indeed, we are all better off when we take advantage of people's unique contributions to our world.

If I have learned anything from a decade and a half of research, in dozens of different countries, it is this: It is much easier to be an optimist in a world that turns people's vulnerabilities into strengths.

Appendix

Exercise One: 12 Things We Need to Succeed

I AM OFTEN ASKED by people who come to hear me speak, "How do I know if I am resilient?" The better question, of course, is, "Do I have the resources to succeed when bad things happen?" While there is no perfect way to assess your resilience (indeed, there are dozens of possible measures, including one of my own called the Adult Resilience Measure), there are some questions we might ask ourselves to quickly audit our environment. Here are a few of my favorites. There are no right answers, but in general, I have learned that the more of these sentences I can complete, the better I will do when stressed.

12 Things We Need to Succeed

1. Structure:

   "There are people in my life who expect me to _____."

2. Consequences:

   "When I don't meet expectations, I know that _____ will happen."

3. Intimate relationships:

   "I can reach out to my _____ to get help when I need it."

4. Other relationships:

   "When bad things happen in my life, there are people like _____ who will support me as best they can."

5. Identity:

   "I feel respected for what is special about me when I'm with/ at/doing _____."

6. Power and control:

   "In my _____ I get to participate in making decisions that affect my _____."

7. Belonging, spirituality, sense of culture:

   "At my _____ people miss me when I'm not there."

   "There are places such as _____ where I can celebrate my culture and beliefs."

8. Rights and responsibilities (social justice):

   "When I'm with others at my _____ I feel treated fairly."

   "When I'm with _____ I am responsible for myself/ others."

9. Safety and security:

   "I am well-cared for by _____."

   "I feel safe when I'm with/at _____."

10. Positive thinking:

    "Though I have problems, I know that things will get better when _____."

    "I know there are good things about me, such as _____."

11. Physical well-being:

    "I am healthy enough to _____."

12. Financial well-being:

    "I have enough money to _____."

These 12 resilience resources are our shock absorbers when our lives hit a speed bump. They are all closely related to things in our environments that we depend on for our personal well-being. Personal traits are still important, but the more difficult our lives become at home, at work, and in our communities, the more all 12 of these resources matter to achieving success.

Exercise Two: Matching Resilience Resources to Life Challenges

HERE IS A MATCHING EXERCISE you can use to assess whether you have the right resources to tackle the specific types of adversity you are experiencing. The best-matched resources will have a differentially large impact on your likelihood of successfully coping. As you work through the exercise, be sure to consider all 12 resources from Chapter 1. Rank the resources. Remember, the severity and duration of adversity both influence which solutions are likely best for which problems. There may be other personal or collective resources you want to add to the list.

This exercise has three steps:

Step 1: Think about a time in your life when you were under unusual stress but managed to cope. Describe the resilience resources you used. In the table provided, tick all those that applied, or add others that are not mentioned.

Step 2: Rank the resources from the most to the least important (I suggest using the number 1 to indicate the most important and number 12, or a higher number, to indicate the least useful to solving this particular problem).

Step 3: Who and/or what was required to make these resources available and accessible to you?

Table 1

Resilience resource *(add your own in the spaces provided)*	Resource was used? *(Yes/No)*	Rank *(1 = most important; 12 or higher = least important)*	Who/what helped make this resource available/accessible?
Intimate relationships			
Lots of strong relationships			
Structure/ routine			
Consequences/ accountability			
A powerful identity			

Resilience resource *(add your own in the spaces provided)*	Resource was used? *(Yes/No)*	Rank *(1 = most important; 12 or higher = least important)*	Who/what helped make this resource available/accessible?
A sense of control			
A sense of belonging/ culture			
Rights and responsibilities			
Safety and support			
Positive thinking			
Physical well-being			

Resilience resource *(add your own in the spaces provided)*	Resource was used? *(Yes/No)*	Rank *(1 = most important; 12 or higher = least important)*	Who/what helped make this resource available/accessible?
Financial well-being			
Other resilience resources?			
•			
•			
•			
•			
•			

Notes

1 Ungar, M., Liebenberg, L., Armstrong, M., Dudding, P., & van de Vijver, F.J.R. (2012). Patterns of service use, individual and contextual risk factors, and resilience among adolescents using multiple psychosocial services. Child Abuse & Neglect, 37(2–3), 150–159.
2 Centers for Disease Control and Prevention. (2017). National marriage and divorce rate trends.
3 Hales, C. M., Carroll, M. D., Fryar, C. D., & Ogden, C. L. (2017). Prevalence of obesity among adults and youth: United States, 2015–2016. NCHS Data Brief No. 288; National Institute of Diabetes and Digestive and Kidney Diseases (NIDDK). (2017). Overweight & obesity statistics.
4 Navaneelan, T., & Janz, T. (2014). Adjusting the scales: Obesity in the Canadian population after correcting for respondent bias. Statistics Canada catalogue no. 82-624-X.
5 Twells, L. K., Gregory, D., Raddigan, J., & Midodzi, W. K. (2014). Current and predicted prevalence of obesity in Canada: A trend analysis. CMAJ Open, 2(1), E18–E26.
6 Statistics Canada. (2015a). Overweight and obese adults (self-reported), 2014."
7 Public Health Agency of Canada. (2016). Report from the Canadian chronic disease surveillance system: Mood and anxiety disorders in Canada, 2016. Ottawa: Public Health Agency of Canada.
8 McDermott, K. W., Elixhauser, A., & Sun, R. (2017). Trends in hospital inpatient stays in the United States, 2005–2014. Healthcare Cost and Utilization Project Statistical Brief #225.
9 Centers for Disease Control and Prevention. (2018). Selected prescription drug classes used in the past 30 days, by sex and age: United States, selected years 1988–1994 through 2011–2014; Jorm, A. F., Patten, S. B., Brugha, T. S., & Majtabai, R. (2017). Has increased provision

of treatment reduced the prevalence of common mental disorders? Review of the evidence from four countries. World Psychiatry, 16, 90–99.

10 Jonas, B. S., Gu, Q., & Albertorio-Diaz, J. R. (2013). Psychotropic medication use among adolescents: United States, 2005–2010. NCHS Data Brief No. 135.

11 Public Health Agency of Canada. (2016). Report from the Canadian chronic disease surveillance system: Mood and anxiety disorders in Canada, 2016. Ottawa: Public Health Agency of Canada.

12 Hemels, M. E. H., Koren, G., & Einarson, T. R. (2002). Increased Use of Antidepressants in Canada: 1981–2000. The Annals of Pharmacotherapy, 36, 1375–1379.

13 Statistics Canada. (2012). Work absence rates. Catalogue no. 71-211-X.

14 Greene, B., Miller, R., H., Crowson, H., Duke, B., & Akey, K. (2004). Predicting high school students' cognitive engagement and achievement: Contributions of classroom perceptions and motivation. Contemporary Educational Psychology, 29(4), 462–482.

15 Gladwell, M. (2008). Outliers: The story of success. New York, NY: Little, Brown.

16 To assess how many of the twelve resilience resources you have and which are most useful to you, complete the exercises in the Appendix.

17 Hobfoll, S. E., & de Jong, T. V. M. (2014). Sociocultural and ecological views of trauma: Replacing cognitive-emotional models of trauma. In L. A. Zoellner & N. C. Feeny (Eds.), *Facilitating resilience and recovery following trauma* (pp. 69–90). New York, NY: Guilford.

18 Wiseman, R. (2003). *The luck factor: Changing your luck, changing your life*. New York, NY: Hyperion.

19 Peterson, C., Park, N., Pole, N., D'Adrea, W., & Seligman, M. E. P. (2008). Strengths of character and posttraumatic growth. *Journal of Traumatic Stress*, 21(2), 214–217.

20 Chen, X., Cen, G., Li, D., & He, Y. (2005). Social functioning and adjustment in Chinese children: The imprint of historical time. *Child Development*, 76(1), 185–192; Chen, X., & Rubin, K. H. (2011). Culture and socioemotional development. In X. Chen & K. H. Rubin (Eds.), *Socioemotional development in cultural context* (pp. 1–8). New York, NY: Guilford.

21 McCubbin, H. I., & Patterson, J. M. (1983). The family stress process: The Double ABCX model of adjustment and adaptation. *Marriage and Family Review*, 6, 7–37.

22 Hobfollo, S. E., & de Jong, T. V. M. (2014). Sociocultural and ecological views of trauma: Replacing cognitive-emotional models of trauma. In L. A. Zoellner & N. C. Feeny (Eds.), *Facilitating resilience and recovery following trauma* (pp. 69–90). New York, NY: Guilford; McCubbin, M. A., & McCubbin, H. I. (1987). Family stress theory and assessment: The T-Double ABCX Model of family adjustment and adaptation. In H. I. McCubbin & A. Thompson (Eds.), *Family assessment inventories for research and practice* (pp. 3–32). Madison, WI: University of Wisconsin—Madison.

23 Ungar, M. (2007). *Too safe for their own good: How risk and responsibility help teens thrive.* Toronto, ON: McClelland & Stewart.

24 Canadian Institute for Health Information. (2015). *Care for children and youth with mental disorders: Report.*

25 McDermott, K. W., Elixhauser, A., & Sun, R. (2017). *Trends in hospital inpatient stays in the United States, 2005–2014.* Healthcare Cost and Utilization Project: Statistical Brief #225; Sun, R., Karaca, Z., & Wong, H. S. (2018). *Trends in hospital emergency department visits by age and payer, 2006–2015.* Healthcare Cost and Utilization Project: Statistical Brief #238. Rockville, MD: Agency for Healthcare Research and Quality.

26 Ungar, M. (2018). The differential impact of social services on young people's resilience. *Child Abuse & Neglect, 78,* 4–13.

27 Zoellner, L. A., & Feeny, N. C. (2014). Conceptualizing risk and resilience following trauma exposure. In L. A. Zoellner & N. C. Feeny (Eds.), *Facilitating resilience and recovery following trauma* (pp. 3–14). New York, NY: Guilford.

28 Van Ijzendoorn, M. H., & Bakermans-Kranenburg, M. J. (2015). Genetic differential susceptibility on trial: Meta-analytic support from randomized controlled experiments. *Development and Psychopathology, 27,* 151–162.

29 Cicchetti, D., Toth, S. L., & Handley, E. D. (2015). Genetic moderation of interpersonal psychotherapy efficacy for low-income mothers with major depressive disorder: Implications for differential susceptibility. *Development and Psychopathology, 27,* 19–35.

30 Davis, E., Keller, J., Hallmayer, J., Ryan, H., Murphy, G., Gotlib, I., & Schatzberg, A. (2017). Association of CRHR1 TAT haplotype with cognitive features of major depressive disorder. *Biological Psychiatry, 81*(10), S225–S226.

31 Ungar, M. (2015). Patterns of family resilience. *Journal of Marital and Family Therapy, 42*(1), 19–31.

32 Masten, A. S., & Wright, M. O. (2010). Resilience over the lifespan: Developmental perspectives on resistance, recovery, and transformation. In J. W. Reich, A. J. Zautra, & J. S. Hall (Eds.), *Handbook of adult resilience* (pp. 213–237). New York, NY: Guilford.

33 Mustanksi, B., & Liu, R. T. (2013). A longitudinal study of predictors of suicide attempts among lesbian, gay, bisexual, and transgender youth. *Archives of Sexual Behavior, 42,* 437–448.

34 Jenny, A., & Alaggia, R. (2012). Children's exposure to domestic violence: Integrating policy, research, and practice to address children's mental health. In R. Alaggia & C. Vine (Eds.), *Cruel but not unusual: Violence in Canadian families* (2nd ed., pp.303–336). Waterloo: Wilfrid University Press.

35 Baskin, C. (2012). Systemic oppression, violence, and healing in Aboriginal families and communities. In R. Alaggia & C. Vine (Eds.), *Cruel but not unusual: Violence in Canadian families* (2nd ed., pp. 147–178). Waterloo, ON: Wilfrid University Press; Ledogar, R. J., & Fleming, J. (2008). Social capital and resilience: A review of concepts and selected literature relevant to Aboriginal youth. *Pimatisiwin: A Journal of Aboriginal and Indigenous Community Health, 6*(2), 25–46.

36 Walsh, F. (2012). *Normal family processes* (4th ed). New York, NY: Guilford.

37 Bonanno, G. A., & Diminich, E. D. (2013). Annual research review: Positive adjustment to adversity—trajectories of minimal-impact resilience and emergent resilience. *Journal of Child Psychology and Psychiatry, 54*(4), 378–401.

38 Bonanno, G. A., & Mancini, A. D. (2012). Beyond resilience and PTSD: Mapping the heterogeneity of responses to potential trauma. *Psychological Trauma, 4*(1), 74–83.

39 Tedeschi, R. G., & Calhoun, L. G. (2004). Posttraumatic growth: Conceptual foundations and empirical evidence. *Psychological Inquiry, 15*(1), 1–18.

40 Davies, K., & Honeyman, G. (2013). Living with a child whose behaviour is described as challenging. *Advances in Mental Health and Intellectual Disabilities, 7*(2), 117–123.

41 Anderson, J. R., Amanor-Boadu, Y., Stith, S. M., & Foster, R. E. (2013). Resilience in military marriages experiencing deployment. In D. Becvar (Ed.), *Handbook of family resilience* (pp. 105–118). New York, NY: Springer.

42 Vaillant, G. E. (2015). Resilience and posttraumatic growth. In D. V. Jeste & B. W. Palmer (Eds.), *Positive psychiatry: A clinical*

handbook (pp. 45–70). Washington, DC: American Psychiatric Publishing (p. 47).

43 Obradović, J., Bush, N. R., Stamperdahl, J., Adler, N. E., & Boyce, W. T. (2010). Biological sensitivity to context: The interactive effects of stress reactivity and family adversity on socioemotional behavior and school readiness. *Child Development, 81*(1), 270–289.

44 Van Voorhees, E. E., Dedert, E. A., Calhoun, P. S., Brancu, M., Runals, J., & VA Mid-Atlantic MIRECC Workgroup. (2012). Childhood trauma exposure in Iraq and Afghanistan war era veterans: Implications for posttraumatic stress disorder symptoms and adult functional social support. *Child Abuse & Neglect, 36*(5), 423–432.

45 Grant, B. F., Stinson, F. S., Hasin, D. S., Dawson, D. A., Chou S. P., Ruan, W. J., & Huang, B. (2004). Immigration and lifetime prevalence of DSM-IV psychiatric disorders among Mexican Americans and non-Hispanic Whites in the United States: Results from the national Epidemiologic Survey on Alcohol and Related Conditions. *Archives of General Psychiatry, 61*(12), 1226–1233.

46 Ungar, M. (2004). *Nurturing hidden resilience in troubled youth.* Toronto, ON: University of Toronto Press.

47 Ziervogel, C. F., Ahmed, N., Fisher, A. J., & Robertson, B. A. (1997). Alcohol misuse in South African male adolescents: A qualitative investigation. *International Quarterly of Community Health Education, 17*(1), 25–41.

48 Flanagan, P. (1998). Teen mothers: Countering the myths of dysfunction and developmental disruption. In C. G. Coll, J. L. Surrey, & K. Weingarten (Eds.), *Mothering against the odds: Diverse voices of contemporary mothers* (pp. 238–254). New York, NY: Guilford.

49 Tolle, E. (1997). *The power of now.* Novato, CA: New World.

50 Garg, E., Chen, L., Nguyen, T., Pokhvisneva, I., Chen, L., Unternaehrer, E., . . . O'Donnell, K. (2018). The early care environment and DNA methylome variation in childhood. *Development and Psychopathology, 30*(3), 891–903.

51 Kalisch, R., Müller, M. B., & Tüsher, O. (2015). A conceptual framework for the neurobiological study of resilience. *Behavioral and Brain Science, 38*, 1–21.

52 Szyf, M., & Pluess, M. (2015). Epigenetics and well-being: optimal adaptation to the environment. In M. Pluess (Ed.), *Genetics of psychological well-being: The role of heritability and genes in positive psychology* (pp. 211–229). Oxford: Oxford Scholarship Online.

53 McDade, T. W., Hoke, M., Borja, J. B., Adair, L. S., & Kuzawa, C. (2013). Do environments in infancy moderate the association between

stress and inflammation in adulthood? Initial evidence from a birth cohort in the Philippines. *Brain, Behavior, and Immunity, 31,* 23–30.

54 Abramson, D. M., Park, Y. S., Stehling-Ariza, T., & Redlener, I. (2010). Children as bellwethers of recovery: Dysfunctional systems and the effects of parents, households, and neighborhoods on serious emotional disturbance in children after Hurricane Katrina. *Disaster Medicine and Public Health Preparedness, 4*(S1), 17–27.

55 Anda, R. F., Felitti, V. J., Bremner, J. D., Walker, J. D., Whitfield, C., Perry, B. D., . . . Giles, W. H. (2006). The enduring effects of abuse and related adverse experiences in childhood: A convergence of evidence from neurobiology and epidemiology. *European Archives of Psychiatry and Clinical Neuroscience, 256,* 174–186.

56 Watamura, S. E., Phillips, D. A., Morrissey, T. W., McCartney, K., & Bub, K. (2011). Double jeopardy: Poorer social-emotional outcomes for children in the NICHD SECCYD experiencing home and child-care environments that confer risk. *Child Development, 82*(1), 48–65.

57 Belsky, D. W., Caspi, A., Cohen, H. J., Kraus, W. E., Ramrakha, S., Poulton, R., Moffitt, T. E. (2017). Impact of early personal-history characteristics on the Pace of Aging: implications for clinical trials of therapies to slow aging and extend healthspan. *Aging Cell, 16*(4), 644–651.

58 Anda, R. F., Felitti, V. J., Bremner, J. D., Walker, J. D., Whitfield, C., Perry, B. D., . . . Giles, W. H. (2006). The enduring effects of abuse and related adverse experiences in childhood: A convergence of evidence from neurobiology and epidemiology. *European Archives of Psychiatry and Clinical Neuroscience, 256,* 174–186.

59 Larkin, H., Shields, J., & Anda, R. (2012). The health and social consequences of Adverse Childhood Experiences (ACE) across the lifespan: An introduction to prevention and intervention in the community. *Journal of Prevention & Intervention in the Community, 40*(4), 263–270.

60 Morris, A. S. (2017). Assessing resilience by examining both adverse and protective childhood experiences. Presentation at biennial meeting of the Society for Research on Child Development, Austin, TX.

61 Heim, C., & Binder, E. B. (2012). Current research trends in early life stress and depression: Review of human studies on sensitive periods, gene-environment interactions, and epigenetics. *Experimental Neurology, 233,* 102–111; Van IJzendoorn, M. H., & Bakermans-Kranenburg, M. J. (2015). Genetic differential susceptibility on trial:

Meta-analytic support from randomized controlled experiments. *Development and Psychopathology, 27,* 151–162.

62 Tetlock, P. E., & Gardner, D. (2015, December 26). Doctors without science. *The Walrus.*

63 Kabat-Zinn, J. (1990). *Full catastrophe living: The program of the stress reduction clinic at the University of Massachusetts Medical Center.* New York, NY: Delta.

64 Perry, B. D. (2009). Examining child maltreatment through a neurodevelopmental lens: Clinical application of the Neurosequential Model of Therapeutics. *Journal of Loss and Trauma, 14,* 240–255.

65 Miller, A. L., Rathus, J. H., & Linehan, M. M. (2007). *Dialectical behavior therapy with suicidal adolescents.* New York, NY: Guilford.

66 Goldberg, S. B., Tucker, R. P., Greene, P. A., Davidson, R. J., Wampold, B. E., Kearney, D. J., & Simpson, T. L. (2018). Mindfulness-based interventions for psychiatric disorders: A systematic review and meta-analysis. *Clinical Psychology Review, 59,* 52–60.

67 Churchill, R., Moore, T. H. M., Furukawa, T. A., Caldwell, D. M., Jones, D. P., Shinohara, K., . . . Hunot, V. (2013). "Third wave" cognitive and behavioural therapies versus treatment as usual for depression. *Cochrane Database of Systematic Reviews,* 10.

68 Cramer, H., Lauche, R., Paul, A., & Dobos, G. (2012). Mindfulness-based stress reduction for breast cancer—a systematic review and meta-analysis. *Current Oncology, 19(5),* e343–e352.

69 Martela, F., & Steger, M. F. (2016). The three meanings of meaning in life: Distinguishing coherence, purpose, and significance. *The Journal of Positive Psychology, 11(5),* 531–545.

70 Fomby, P., & Cherlin, A. J. (2007). Family instability and child well-being. *American Sociological Review, 72(2),* 181–204.

71 Osborne, C., & McLanahan, S. (2007). Partnership instability and child well-being. *Journal of Marriage and Family, 69,* 1065–1083.

72 Cuddy, A. (2015). *Presence: Bringing your boldest self to your biggest challenges.* New York, NY: Little, Brown.

73 Levy, B. R., Slade, M. D., Kunkel, S., & Kasl, S. (2002). Longevity increased by positive self-perceptions of aging. *Journal of Personality & Social Psychology, 83,* 261–270.

74 Levy, B. R., & Myers, L. M. (2004). Preventive health behaviors influenced by self-perceptions of aging. *Preventive Medicine, 39,* 625–629; Meisner, B. A., & Baker, J. (2013). An exploratory analysis of aging expectations and health care behavior among aging adults. *Psychology & Aging, 28,* 99–104.

75 Sarkisian, C.A., Hays, R. D., & Mangioine, C. M. (2002). Do older adults expect to age successfully? The association between expectations regarding aging and beliefs regarding healthcare seeking among older adults. *Journal of the American Geriatric Society, 50,* 1837–1843.

76 Jeste, D. V., Savla, G. N., Thompson, W. K., Vahia, I. V., Glorioso, D. K., Martin, A. S., . . . Depp, C. A. (2013). Association between older age and more successful aging: Critical role of resilience and depression. *The American Journal of Psychiatry, 170*(2), 188–196.

77 Ehrenreich, B. (2009). *Bright-sided: How the relentless promotion of positive thinking has undermined America.* New York, NY: Metropolitan.

78 Leighton, D. C., Harding, J. S., Macklin, D. B., Hughes, C. C., & Leighton, A. H. (1963). Psychiatric findings of the Stirling Country Study. *American Journal of Psychiatry, 119,* 1021–1026.

79 Botey, A. P., & Kulig, J. C. (2013). Family functioning following wildfires: Recovering from the 2011 Slave Lake fires. *Journal of Child and Family Studies, 23*(8), 1471–1483.

80 Botey, A. P., & Kulig, J. C. (2013). Family functioning following wildfires: Recovering from the 2011 Slave Lake fires. *Journal of Child and Family Studies, 23*(8), 1471–1483.

81 Cox, R. S. (2015). *Measuring community disaster resilience: A review of current theories and practices with recommendations.* Ottawa, ON: International Safety Research.

82 Bandura, A. (1998). Exercise of agency in personal and social change. In E. Sanavio (Ed.), *Behaviour and cognitive therapy today: Essays in honor of Hans J. Eysenck* (pp. 1–29). Oxford, UK: Pergamon.

83 Wiesel, E. (1956). *Night.* New York, NY: Hill and Wang.

84 Werner, E. E., & Smith, R. S. (2001). *Journeys from childhood to midlife: Risk, resilience, and recovery.* Ithaca, NY: Cornell University Press; Johnson, S. (2008). *Hold me tight: Seven conversations for a lifetime of love.* New York, NY: Little, Brown.

85 Beckett, C., Maughan, B., Rutter, M., Castle, J., Colvert, E., Groothues, C., . . . Sonuga-Barke, E. J. S. (2006). Do the effects of early severe deprivation on cognition persist into early adolescence? Findings from the English and Romanian Adoptees Study. *Child Development, 77*(3), 696–711.

86 Sampson, R. J., & Laub, J. H. (1997). A life course theory of cumulative disadvantage and the stability of delinquency. In T. P. Thornberry (Ed.), *Developmental theories of crime and delinquency* (pp. 133–161). New Brunswick, NJ: Transaction Publishers.

87 Love, P., & Stosny, S. (2007). *How to improve your marriage without talking about it.* New York, NY: Broadway.

88 Falicov, C. J. (2007). Working with transnational immigrants: Expanding meanings of family, community, and culture. *Family Process, 46*, 157–171; McGoldrick, M. (2003). Culture: A challenge to concepts of normality. In F. Walsh (Ed.), *Normal family processes* (3rd ed., pp. 61–95). New York, NY: Guilford.

89 Ungar, M., Liebenberg, L., Landry, N., & Ikeda, J. (2012). Caregivers, young people with complex needs, and multiple service providers: A study of triangulated relationships and their impact on resilience. *Family Process, 51*(2), 193–206; Weihs, K., Fisher, L., & Baird, M. A. (2002). Families, health and behavior. *Families, Systems & Health, 20*(1), 7–46.

90 Weine, S. M., Levin, E., Hakizimana, L., & Kahnweih, G. (2012). How prior social ecologies shape family resilience amongst refugees in U.S. resettlement. In M. Ungar (Ed.), *The social ecology of resilience: A handbook of theory and practice* (pp. 309–324). New York, NY: Springer.

91 Gaugler, J. E. (2010). The longitudinal ramifications of stroke caregiving: A systematic review. *Rehabilitation Psychology, 55*(2),108–125; McCubbin, L. D., & McCubbin, H. I. (2013). Resilience in ethnic family systems: A relational theory for research and practice. In D. Becvar (Ed.), *Handbook of family resilience* (pp. 175–195). New York, NY: Springer.

92 McAdams, D. (1993). *The stories we live by*. New York, NY: William Morrow.

93 Winwood, P. C., Colon, R., & McEwan, K. (2013). A practical measure of workplace resilience: Developing the resilience at work scale. *Journal of Occupational and Environmental Management, 55*(10), 1205–1212.

94 Campbell, C., Ungar, M., & Dutton, P. (2008). *The decade after high school: A parent's guide*. Halifax, NS: Canadian Education and Research Institute for Counseling and the Resilience Research Centre.

95 Hafez, M., & Mullins, C. (2015). The radicalization puzzle: A theoretical synthesis of empirical approaches to homegrown extremism. *Studies in Conflict and Terrorism, 38*(11), 958–975.

96 Moghaddam, F. M. (2005). The staircase to terrorism. *American Psychologist, 60*(2), 161–169.

97 Gonzalez, K. A., Rostosky, S. S., Odom, R. D., & Riggle, E. D. B. (2013). The positive aspects of being the parent of an LGBTQ child. *Family Process, 52*(2), 325–337; LaSala, M. C. (2010). *Coming out, coming home: Helping families adjust to a gay or lesbian child*. New York, NY: Columbia University Press.

98 Werner, E. E., & Smith, R. S. (2001). *Journeys from childhood to midlife: Risk, resilience, and recovery*. Ithaca, NY: Cornell University Press.

99 Borum, R. (2011). Radicalization into violent extremism I: A review of social science theories. *Journal of Strategic Security, 4*(4), 2.

100 Zinger, D. (2016, March 21). How to engage employees. Power of Happiness Annual Conference, Istanbul.

101 Bonanno, G. A., & Diminich, E. D. (2013). Annual research review: Positive adjustment to adversity—trajectories of minimal-impact resilience and emergent resilience. *Journal of Child Psychology and Psychiatry, 54*(4), 378–401; Larkin, H., Shields, J. J., & Anda, R. F. (2012). The health and social consequences of adverse childhood experiences (ACE) across the lifespan: An introduction to prevention and intervention in the community. *Journal of Prevention & Intervention in the Community, 40,* 263–270.

102 Bonanno, G. A., & Mancini, A. D. (2012). Beyond resilience and PTSD: Mapping the heterogeneity of responses to potential trauma. *Psychological Trauma, 4*(1), 74–83.

103 Borum, R. (2011). Radicalization into violent extremism I: A review of social science theories. *Journal of Strategic Security, 4*(4), 2; Hafez, M., & Mullins, C. (2015). The radicalization puzzle: A theoretical synthesis of empirical approaches to homegrown extremism. *Studies in Conflict and Terrorism, 38*(11), 958–975; Harris-Hogan, S., Barrelle, K., & Zammit, A. (2016). What is countering violent extremism? Exploring CVE policy and practice in Australia. *Behavioral Sciences of Terrorism and Political Aggression, 8*(1), 6–24; Khalil, J., & Zeuthen, M. (2016). *Countering violent extremism and risk reduction: A guide to programme design and evaluation*. Whitehall Report 2-16. London: Royal United Services Institute for Defence and Security Studies; Weine, S. (2017). Resilience and countering violent extremism. In U. Kumar (Ed.), *The Routledge international handbook of psychosocial resilience* (pp. 189–201). London: Routledge.

104 Miller, S, Hubble, M., & Mathieu, F. (2015). Burnout reconsidered: What supershrinks can teach us. *Psychotherapy Networker, 39*(3), 18–23.

105 Biswas-Diener, R. (Ed.). (2011). *Positive psychology as social change.* Springer Netherlands.

106 Goetzel, R. Z., & Ozminkowski, R. J. (2008). The health and cost benefits of work site health-promotion programs. *Annual Review of Public Health, 29*(1), 303–323.

107 Ellis, B. J., & Boyce, W. T. (2011). Differential susceptibility to the environment: Toward an understanding of sensitivity to develop-

mental experiences and context. *Development and Psychopathology, 23,* 1–5.

108 Ellis, B. J. & Boyce. W. T. (2011). Differential susceptibility to the environment: Toward an understanding of sensitivity to developmental experiences and context. *Development and Psychopathology, 23,* 1–5.

109 Pauly, D. (2009, September 28). Aquacalypse now: The end of fish. *New Republic.*

110 Béné, C. (2003). When fishery rhymes with poverty: A first step beyond the old paradigm on poverty in small-scale fisheries. *World Development, 31*(6), 949–975.

111 Campbell, C., & Ungar, M. (2008, July). *The decade after high school: A professional's guide.* Halifax, NS: Education and Research Institute for Counseling, and the Resilience Research Centre; Campbell, C. G., & Ungar, M. (2004). Constructing a life that works: Part One. The fit between postmodern family therapy and career counselling. *The Career Development Quarterly, 53*(1), 16–27.

112 Betsworth, D. G., & Hansen, J. C. (1996). The categorization of serendipitous career development events. *Journal of Career Assessment, 4*(1), 91–98.

113 Koonce, R. (1995) Becoming your own career coach. *Training and Development, 49*(1), 18–25.

114 Boynton, A. (2011, October 18). Are you an "I" or a "T"? *Forbes.*

115 London, M., & Noe, R. A. (1997). London's career motivation theory: An update on measurement and research. *Journal of Career Assessment, 5,* 61–80 (p. 62).

116 Bridges, W. (1994). *Job Shift.* Reading, MA: Addison-Wesley (p. 57).

117 Schoon, I. (2006). *Risk and resilience: Adaptations in changing times.* Cambridge: Cambridge University Press.

118 Keravala, J. (2018). Technology is changing the world for good. IFC Sustainability Exchange, 2018. Washington, DC.

119 Seba, T. (2014). *Clean disruption of energy and transportation.* Silicon Valley, CA: Clean Planet Ventures.

120 A shorter list of principles was first published in the following documents: Campbell, C. G., & Ungar, M. (2004). Constructing a life that works: Part two. An approach to practice. *The Career Development Quarterly, 53*(1), 28–40; Campbell, C. G., & Ungar, M. (2008). The decade after high school: Professionals guide. Ottawa, ON: CERIC; Campbell, C. G. (2012). *A study of the career pathways of Canadian young adults during the decade after secondary school graduation* (Doctoral thesis). Massey University, New Zealand.

121 Sher, B., & Gottleib, A. (1979). *Wishcraft*. New York, NY: Ballantine (p.xiv).

122 Aiken, S. (2010). *The one-week job project*. Toronto, ON: Penguin.

123 Grant, T. (2009, May 30). Democracy in action. *Globe and Mail*.

124 Montana, P. J., & Lenaghan, J. A. (1999). What motivates and matters most to Generations X and Y. *Journal of Career Planning & Employment*, 59(4), 27–30.

125 Amudson, N. (2003). *The physics of living*. Richmond, BC: Ergon Communications.

126 Thaler, R. H., & Sunstein, C. R. (2009). *Nudge: Improving decisions about health, wealth, and happiness*. London: Penguin Books.

127 Seligman, M. E. P. (2011). *Flourish*. New York, NY: Free Press.

128 Walsh, F. (2006). *Strengthening family resilience*. New York, NY: Guilford.

129 Canadian Institute for Health Information. (2015). *Care for children and youth with mental disorders: Report*.

130 Bonanno, G. A., & Mancini, A. D. (2012). Beyond resilience and PTSD: Mapping the heterogeneity of responses to potential trauma. *Psychological Trauma*, 4(1), 74–83.

131 Boyce, W. T., Sokolowski, M. B., & Robinson, G. E. (2012). Toward a new biology of social adversity. *Proceedings of the National Academy of Sciences of the United States of America*, 109(S2), 17143–17148.

132 Real, T. (2007). *The new rules of marriage*. New York, NY: Ballantine.

133 Ungar, M. (2015). Social ecological complexity and resilience processes. Commentary on "A conceptual framework for the neurobiological study of resilience." *Behavioral and Brain Sciences*, 38, 50–51.

134 O'Brien, G., & Hope, A. (2010). Localism and energy: Negotiating approaches to embedding resilience in energy systems. *Energy Policy*, 38(12), 7550–7558.

135 McKenzie-Mohr, S. & Lafrance, M. (2011). Telling stories without the words: "Tightrope talk" in women's accounts of coming to live well after rape or depression. *Feminism & Psychology*, 21(1), 49–73.

136 Byrne, R. (2006). *The secret*. New York, NY: Atria.

137 Heemskerk, F. (2018, May 24). How to strive and thrive in the future economy. IFC Sustainability Exchange, Washington, DC.

138 Chung, R. C., & Bemak, F. P. (2012). *Social justice counselling: The next steps beyond multiculturalism*. Thousand Oaks, CA: Sage; Fisher, C. B., Busch-Rossnagel, N. A., Jopp, D. S., & Brown, J. L. (2012). Applied developmental science, social justice, and socio-political well-being. *Applied Developmental Science*, 16(1), 54–64.

139 Murphy, D. (2017). *Health insurance coverage improves child well-being*. Child Trends Research Brief. Publication #2017–22.

140 Canadian Institute for Health Information. (2017). *Wait times for priority procedures in Canada, 2017.* Ottawa, ON: CIHI.

141 Boivin, M., Hertzman, C., Barr, R., Boyce, T. Fleming, A., MacMillan, H., . . . Trocmé, N. (2013). *Early childhood development: Adverse experiences and developmental health.* Ottawa, ON: Royal Society of Canada—Canadian Academy of Health Sciences Expert Panel.

142 Abramson, D. M., & Garfield, R. M. (2006). *On the edge: Children and families displaced by Hurricanes Katrina and Rita face a looming medical and mental health crisis.* New York, NY: Columbia University, National Center for Disaster Preparedness.

143 See, for example, Ungar, M. (2011). The social ecology of resilience: Addressing contextual and cultural ambiguity of a nascent construct. *American Journal of Orthopsychiatry, 81*(1), 1–17; Werner, E. E. (2005). What can we learn about resilience from large-scale longitudinal studies? In S. Goldstein & R. B. Brooks (Eds.), *Handbook of resilience in children* (pp. 91–105). New York, NY: Kluwer Academic/Plenum Publishers.

144 Duncan, B. L., Miller, S. D., & Sparks, J. A. (2004). *The heroic client.* New York, NY: Jossey-Bass.

145 Grossman, M., Ungar, M., & Amarasingam, A. (2015). *Indicators for building resilience against violent extremism in culturally diverse communities.* Australia/New Zealand: Counter-Terrorism Committee/Attorney-General's Department.

146 Harvey, J., & Delfabbro, P. H. (2004). Psychological resilience in disadvantaged youth: A critical overview. *Australian Psychologist, 39*(1), 3–13; Norris, F. H., Sherrieb, K., & Pfefferbaum, B. (2011). Community resilience: Concepts, assessment, and implications for intervention. In S. M. Southwick, B. T. Litz, D. Charney, & M. J. Friedman (Eds.), *Resilience and mental health: Challenges across the lifespan* (pp. 162–175). New York: Cambridge University Press.

147 Maslow, A. H. (1943). A theory of human motivation. *Psychological Review, 50*(4), 370–396.

148 Sinek, S. (2013). *Leaders eat last.* New York, NY: Penguin.

149 Dudley, D., Cairney, J., Wainwright, N., Kriellaars, D., & Mitchell, D. (2017). Critical considerations for physical literacy policy in public health, recreation, sport, and education agencies. *Quest, 69*(4), 436–452,

150 Gotts, N. M. (2007). Resilience, panarchy, and world-systems analysis. *Ecology and Society, 12*(1), 24; Folke, C., Carpenter, S. R., Walker, B., Scheffer, M., Chapin, T., & Rockström, J. (2010). Resilience thinking:

Integrating resilience, adaptability and transformability. *Ecology and Society, 15*(4), 20; Ungar, M. (2011). The social ecology of resilience: Addressing contextual and cultural ambiguity of a nascent construct. *American Journal of Orthopsychiatry, 81*(1), 1–17; Masten, A. S. (2014). Global perspectives on resilience in children and youth. *Child development, 85*(1), 6–20.

151 Munafo, M. (2016, March 4). Genetic denialism is unhelpful—genes play a role in who we are. *The Guardian.*

152 Frankenhuis, W. E., & Del Giudice, M. (2012). When do adaptive developmental mechanisms yield maladaptive outcomes? *Developmental Psychology, 48*(3), 628–642.

153 Baggio, J. A., Brown, K., & Hellebrandt, D. (2015). Boundary object or bridging concept? A citation network analysis of resilience. *Ecology and Society, 20*(2), 2.

154 Brown, K. (2016). *Resilience, global change and development.* London: Routledge.

155 Xu, L., & Kajikawa, Y. (2017). An integrated framework for resilience research: A systematic review based on citation network analysis. *Sustainability Science, 13*(1), 235–254.

156 Xu, L., & Kajikawa, Y. (2017). An integrated framework for resilience research: A systematic review based on citation network analysis. *Sustainability Science, 13*(1), 235–254.

157 Masten, A. S. (2014). Global perspectives on resilience in children and youth. *Child development, 85*(1), 6–20.

158 Baggio, J. A., BurnSilver, S. B., Arenas, A., Magdanz, J. S., Kofinas, G. P., & De Domenico, M. (2016). Multiplex social ecological network analysis reveals how social changes affect community robustness more than resource depletion. *PNAS, 113*(48), 13708–13713.

159 Kofinas, G., Clark, D., & Hovelsrud, G. K. (2013). Adaptive and transformative capacity. In *Arctic Resilience Interim Report 2013* (pp. 73–93). Stockholm Environment Institute and Stockholm Resilience Centre; Carson, M., & Peterson, G. (Eds). (2016). *Arctic resilience report.* Stockholm Environment Institute and Stockholm Resilience Centre, Stockholm.

160 Holling, C. S. (1973). Resilience and stability of ecological systems. *Annual Review Of Ecology & Systematics, 4*, 1–23.

161 Folke, C. (2006). Resilience: The emergence of a perspective for social–ecological systems analyses. *Global Environmental Change, 16*(3), 253–267.

162 Quinlan, A., Berbés-Blázquez, M., Haider, J. L., & Peterson, G. D. (2015). Measuring and assessing resilience: Broadening understanding

through multiple disciplinary perspectives. *Journal of Applied Ecology,* 53, 677–687.

163 Ungar, M. (in press). Systemic resilience: Principles and processes for a science of change in contexts of adversity. *Ecology and Society.*

164 Cadima, J., Enrico, M., Ferreira, T., Verschueren, K., Leal, T., & Matos, P. M. (2016). Self-regulation in early childhood: The interplay between family risk, temperament and teacher-child interactions. *European Journal of Developmental Psychology,* 13(3), 341–360.

165 Sterbenz, J. P. G., Hutchison, D., Çetinkaya, E. K., Jabbar, A., Rohrer, J. P., Schöller, M., & Smith, P. (2010). Resilience and survivability in communication networks: Strategies, principles, and survey of disciplines. *Computer Networks,* 54(8), 1245–1265.

166 Hordge-Freeman, E. (2015). The color of love: Racial features, stigma, and socialization in Black Brazilian families. Austin, TX: University of Texas Press.

167 Obradović, J. (2012). How can the study of physiological reactivity contribute to our understanding of adversity and resilience processes in development? *Development and Psychopathology,* 24(2), 371–387; Ellis, B. J., & Del Giudice, M. (2014). Beyond allostatic load: Rethinking the role of stress in regulating human development. *Development and Psychopathology,* 26, 1–20.

168 Lupien, S. J., King, S., Meaney, M. J., & McEwen, B. S. (2001). Can poverty get under your skin? Basal cortisol levels and cognitive function in children from low and high socioeconomic status. *Development and Psychopathology,* 13(3), 651–674; Boxer, P., Huesmann, L. R., Dubow, E. F., Landau, S. F., Gvirsman, S. D., Shikaki, K., & Ginges, J. (2013). Exposure to violence across the social ecosystem and the development of aggression: A test of ecological theory in the Israeli-Palestinian conflict. *Child Development,* 84(1), 163–177.

169 Ott, E., & Montgomery, P. (2015). Interventions to improve the economic self-sufficiency and well-being of resettled refugees: A systematic review. *Campbell Systematic Reviews,* 11, 1–53.

170 Hou, F., & Bonikowska, A. (2016). *Educational and labour market outcomes of childhood immigrants by immigration class.* Analytical Studies Branch Research Paper Series 377, Catalogue no. 11F0019M. Ottawa, ON: Statistics Canada.

171 Salat, S., & Bourdic, L. (2012). The resilient city 2. *Journal of Land Use, Mobility and Environment,* 5(2), 55–68.

172 Panter-Brick, C. & Eggerman, M. (2012). Understanding culture, resilience, and mental health: The production of hope. In M. Ungar

(Ed.), *The social ecology of resilience: A handbook of theory and practice* (pp. 369–386). New York: Springer.

173 Gunderson, L. H., Allen, C. R., & Holling, C. S. (2010). *Foundation of ecological resilience.* Washington, DC: Island Press.

174 Taylor, J. E., Filipski, M. J., Alloush, M., Gupta, A., Rojas Valdes, R. I., & Gonzalez-Estrada, E. (2016). Economic impact of refugees. *PNAS, 113*(27), 7449–7453.

175 Bodin, Ö. (2017). Collaborative environmental governance: Achieving collective action in social-ecological systems. *Science, 357*(6352), eaan1114.

176 Anderies, J. M., Janssen, M. A., & Schlager, E. (2016). Institutions and the performance of coupled infrastructure systems. *International Journal of the Commons, 10*(2), 495–516.

177 Vygotsky, L. S. (1978). *Mind in society: The development of higher psychological processes.* M. Cole, V. John-Steiner, S. Scribner, & E. Souberman (Eds.). Cambridge, MA: Harvard University Press.

178 Pahl-Wostl, C. (2009). A conceptual framework for analysing adaptive capacity and multi-level learning processes in resource governance regimes. *Global Environmental Change, 19*(3), 354–365.

179 Liew, J., Cao, Q., Hughes, J. N., & Deutz, M. H. F. (2018). Academic resilience despite early academic adversity: A three-wave longitudinal study on regulation-related resiliency, interpersonal relationships, and achievement in first to third grade. *Early Education and Development, 29*(5), 762–779.

180 Prilleltensky, I. & Prilleltensky, O. (2007). *Promoting well-being: Linking personal, organizational, and community change.* New York, NY: Wiley.

181 Werner, E., & Smith, R. (1998). *Vulnerable but invincible: A longitudinal study of resilient children and youth.* New York, NY: Adams, Bannister, Cox.

182 Brown, T. (2009). *Change by design.* New York, NY: HarperCollins.

183 Roche, D. (2012). *The church of 80% sincerity.* Vancouver, BC: Perigee.

184 Harford, T. (2012). *Adapt.* New York, NY: Abacus.

185 Duckworth, A. (2016). *Grit.* New York, NY: Scribner.

186 Dweck, C. A. (2007). *Mindset: The new psychology of success.* New York, NY: Ballantine.

187 McKnight, J. L., & Kretzman, J. P. (1993). *Building communities from the inside out: A path toward finding and mobilizing a community's assets.* Chicago, IL: Institute for Policy Research.

188 Wilson, R. & Lyons, L. (2013). *Anxious kids, anxious parents: Seven ways to stop the worry cycle and raise courageous and independent children.* Boston, MA: HCI Books.

189 Johnson, S. (2008). *Hold me tight: Seven conversations for a lifetime of love.* New York, NY: Little, Brown.

190 Perel, E. (2007). *Mating in captivity.* New York, NY: HarperCollins.

191 McCubbin, H. I., & Patterson, J. M. (1983). The family stress process: The Double ABCX model of adjustment and adaptation. *Marriage and Family Review, 6,* 7–37.

192 Kennedy, P. (2018, March 14). The secret to a longer life? Don't ask these dead longevity researchers. *New York Times.*

193 Roetert, E. P., Kriellaars, D., Ellenbecker, T. S., & Richardson, C. (2017). Preparing students for a physically literate life. *Journal of Physical Education, Recreation & Dance, 88*(1), 57–62.

194 Bacon, L., & Aphramor, L. (2011). Weight science: Evaluating the evidence for a paradigm shift. *Nutrition Journal, 10*(9).

195 Leavy, J. E., Bull, F. C., Rosenberg, M., & Bauman, A. (2011). Physical activity mass media campaigns and their evaluation: A systematic review of the literature 2003–2010. *Health Education Research, 26*(6), 1060–1085.

196 Montgomery, C. (2013). *Happy city: Transforming our lives through urban design.* Toronto, ON: Penguin.

197 See, for example, Alexiou, A. S. (2006). *Jane Jacobs: Urban visionary.* Toronto, ON: HarperCollins.

198 C3 Collaborating for health. (2012). *The benefits of regular walking for health, well-being and the environment.*

199 Gillen, J. B., Martin, B. J., MacInnis, M. J., Skelly, L. E., & Tarnopolsky, M. A. (2016). Twelve weeks of sprint interval training improves indices of cardiometabolic health similar to traditional endurance training despite a five-fold lower exercise volume and time commitment. *PLOS ONE.*

200 Groger, M. (1985). *Eating awareness training.* New York, NY: Simon & Schuster.

201 MacKinnon, J. B. & Smith, A. (2007). *The 100-mile diet: A year of local eating.* Toronto, ON: Random House.

202 Friedrich, R. R., Schuch, I., & Wagner, M. B. (2012). Effect of interventions on the body mass index of school-age students. *Revista de Saúde Pública, 46*(3).

203 Huffington, A. (2016). *The sleep revolution.* New York, NY: Crown.

204 Hunt, D. & Hait, P. (1990). *The Tao of time.* New York, NY: Simon & Schuster.

205 Chilton, D. (2011). *The wealthy barber returns.* Toronto, ON: Financial Awareness Corporation.

206 Daily Mail Reporter. (2012, December 24). How Americans have HALF the amount of time off of nearly every other nation each year . . . and most are too scared for their jobs to use all of their vacation time. *Daily Mail.*

207 Leung, L. (2016). Uses and gratifications. *The International Encyclopedia of Political Communication.*

208 Vanhove, A. J., Herian, M. N., Harms, P. D., Luthans, F., & DeSimone, J. A. (2014). *Examining psychosocial well-being and performance in isolated, confined, and extreme environments: Final report.* Houston, TX: NASA.

209 McEachan, R. C., Yang, T. C., Roberts, H., Pickett, K. E., Arseneau-Powell, D., Gidlow, . . . Nieuwenhuijsen, M. (2018). Availability, use of, and satisfaction with green space, and children's mental well-being at age 4 years in a multicultural, deprived, urban area: results from the Born in Bradford cohort study. *Lancet Planetary Health, 2,* e244–e254.

About the Author

Dr. Michael Ungar is one of the world's leading experts on resilience. He is the Canada Research Chair in Child, Family, and Community Resilience, a professor of social work at Dalhousie University, and a family therapist. He is the author of fourteen books, 150 scientific papers, and a blog for *Psychology Today*. He has worked with dozens of organizations around the world, including the World Bank, UNESCO, and the Red Cross, and is a recipient of the Canadian Association of Social Workers National Distinguished Service Award.